Niels Bohr: A Very Short Introduction

VERY SHORT INTRODUCTIONS are for anyone wanting a stimulating and accessible way into a new subject. They are written by experts, and have been translated into more than 45 different languages.

The series began in 1995, and now covers a wide variety of topics in every discipline. The VSI library currently contains over 600 volumes—a Very Short Introduction to everything from Psychology and Philosophy of Science to American History and Relativity—and continues to grow in every subject area.

Very Short Introductions available now:

ABOLITIONISM Richard S. Newman
THE ABRAHAMIC RELIGIONS Charles L. Cohen
ACCOUNTING Christopher Nobes
ADAM SMITH Christopher J. Berry
ADOLESCENCE Peter K. Smith
ADVERTISING Winston Fletcher
AESTHETICS Bence Nanay
AFRICAN AMERICAN RELIGION Eddie S. Glaude Jr
AFRICAN HISTORY John Parker and Richard Rathbone
AFRICAN POLITICS Ian Taylor
AFRICAN RELIGIONS Jacob K. Olupona
AGEING Nancy A. Pachana
AGNOSTICISM Robin Le Poidevin
AGRICULTURE Paul Brassley and Richard Soffe
ALEXANDER THE GREAT Hugh Bowden
ALGEBRA Peter M. Higgins
AMERICAN CULTURAL HISTORY Eric Avila
AMERICAN FOREIGN RELATIONS Andrew Preston
AMERICAN HISTORY Paul S. Boyer
AMERICAN IMMIGRATION David A. Gerber
AMERICAN LEGAL HISTORY G. Edward White
AMERICAN NAVAL HISTORY Craig L. Symonds
AMERICAN POLITICAL HISTORY Donald Critchlow
AMERICAN POLITICAL PARTIES AND ELECTIONS L. Sandy Maisel
AMERICAN POLITICS Richard M. Valelly
THE AMERICAN PRESIDENCY Charles O. Jones
THE AMERICAN REVOLUTION Robert J. Allison
AMERICAN SLAVERY Heather Andrea Williams
THE AMERICAN WEST Stephen Aron
AMERICAN WOMEN'S HISTORY Susan Ware
ANAESTHESIA Aidan O'Donnell
ANALYTIC PHILOSOPHY Michael Beaney
ANARCHISM Colin Ward
ANCIENT ASSYRIA Karen Radner
ANCIENT EGYPT Ian Shaw
ANCIENT EGYPTIAN ART AND ARCHITECTURE Christina Riggs
ANCIENT GREECE Paul Cartledge
THE ANCIENT NEAR EAST Amanda H. Podany
ANCIENT PHILOSOPHY Julia Annas
ANCIENT WARFARE Harry Sidebottom
ANGELS David Albert Jones
ANGLICANISM Mark Chapman
THE ANGLO-SAXON AGE John Blair
ANIMAL BEHAVIOUR Tristram D. Wyatt
THE ANIMAL KINGDOM Peter Holland
ANIMAL RIGHTS David DeGrazia

THE ANTARCTIC Klaus Dodds
ANTHROPOCENE Erle C. Ellis
ANTISEMITISM Steven Beller
ANXIETY Daniel Freeman and Jason Freeman
THE APOCRYPHAL GOSPELS Paul Foster
APPLIED MATHEMATICS Alain Goriely
ARCHAEOLOGY Paul Bahn
ARCHITECTURE Andrew Ballantyne
ARISTOCRACY William Doyle
ARISTOTLE Jonathan Barnes
ART HISTORY Dana Arnold
ART THEORY Cynthia Freeland
ARTIFICIAL INTELLIGENCE Margaret A. Boden
ASIAN AMERICAN HISTORY Madeline Y. Hsu
ASTROBIOLOGY David C. Catling
ASTROPHYSICS James Binney
ATHEISM Julian Baggini
THE ATMOSPHERE Paul I. Palmer
AUGUSTINE Henry Chadwick
AUSTRALIA Kenneth Morgan
AUTISM Uta Frith
AUTOBIOGRAPHY Laura Marcus
THE AVANT GARDE David Cottington
THE AZTECS Davíd Carrasco
BABYLONIA Trevor Bryce
BACTERIA Sebastian G. B. Amyes
BANKING John Goddard and John O. S. Wilson
BARTHES Jonathan Culler
THE BEATS David Sterritt
BEAUTY Roger Scruton
BEHAVIOURAL ECONOMICS Michelle Baddeley
BESTSELLERS John Sutherland
THE BIBLE John Riches
BIBLICAL ARCHAEOLOGY Eric H. Cline
BIG DATA Dawn E. Holmes
BIOGRAPHY Hermione Lee
BIOMETRICS Michael Fairhurst
BLACK HOLES Katherine Blundell
BLOOD Chris Cooper
THE BLUES Elijah Wald
THE BODY Chris Shilling
THE BOOK OF COMMON PRAYER Brian Cummings
THE BOOK OF MORMON Terryl Givens
BORDERS Alexander C. Diener and Joshua Hagen
THE BRAIN Michael O'Shea
BRANDING Robert Jones
THE BRICS Andrew F. Cooper
THE BRITISH CONSTITUTION Martin Loughlin
THE BRITISH EMPIRE Ashley Jackson
BRITISH POLITICS Anthony Wright
BUDDHA Michael Carrithers
BUDDHISM Damien Keown
BUDDHIST ETHICS Damien Keown
BYZANTIUM Peter Sarris
C. S. LEWIS James Como
CALVINISM Jon Balserak
CANCER Nicholas James
CAPITALISM James Fulcher
CATHOLICISM Gerald O'Collins
CAUSATION Stephen Mumford and Rani Lill Anjum
THE CELL Terence Allen and Graham Cowling
THE CELTS Barry Cunliffe
CHAOS Leonard Smith
CHARLES DICKENS Jenny Hartley
CHEMISTRY Peter Atkins
CHILD PSYCHOLOGY Usha Goswami
CHILDREN'S LITERATURE Kimberley Reynolds
CHINESE LITERATURE Sabina Knight
CHOICE THEORY Michael Allingham
CHRISTIAN ART Beth Williamson
CHRISTIAN ETHICS D. Stephen Long
CHRISTIANITY Linda Woodhead
CIRCADIAN RHYTHMS Russell Foster and Leon Kreitzman
CITIZENSHIP Richard Bellamy
CIVIL ENGINEERING David Muir Wood
CLASSICAL LITERATURE William Allan
CLASSICAL MYTHOLOGY Helen Morales
CLASSICS Mary Beard and John Henderson
CLAUSEWITZ Michael Howard
CLIMATE Mark Maslin
CLIMATE CHANGE Mark Maslin
CLINICAL PSYCHOLOGY Susan Llewelyn and Katie Aafjes-van Doorn

COGNITIVE NEUROSCIENCE Richard Passingham
THE COLD WAR Robert McMahon
COLONIAL AMERICA Alan Taylor
COLONIAL LATIN AMERICAN LITERATURE Rolena Adorno
COMBINATORICS Robin Wilson
COMEDY Matthew Bevis
COMMUNISM Leslie Holmes
COMPARATIVE LITERATURE Ben Hutchinson
COMPLEXITY John H. Holland
THE COMPUTER Darrel Ince
COMPUTER SCIENCE Subrata Dasgupta
CONCENTRATION CAMPS Dan Stone
CONFUCIANISM Daniel K. Gardner
THE CONQUISTADORS Matthew Restall and Felipe Fernández-Armesto
CONSCIENCE Paul Strohm
CONSCIOUSNESS Susan Blackmore
CONTEMPORARY ART Julian Stallabrass
CONTEMPORARY FICTION Robert Eaglestone
CONTINENTAL PHILOSOPHY Simon Critchley
COPERNICUS Owen Gingerich
CORAL REEFS Charles Sheppard
CORPORATE SOCIAL RESPONSIBILITY Jeremy Moon
CORRUPTION Leslie Holmes
COSMOLOGY Peter Coles
COUNTRY MUSIC Richard Carlin
CRIME FICTION Richard Bradford
CRIMINAL JUSTICE Julian V. Roberts
CRIMINOLOGY Tim Newburn
CRITICAL THEORY Stephen Eric Bronner
THE CRUSADES Christopher Tyerman
CRYPTOGRAPHY Fred Piper and Sean Murphy
CRYSTALLOGRAPHY A. M. Glazer
THE CULTURAL REVOLUTION Richard Curt Kraus
DADA AND SURREALISM David Hopkins
DANTE Peter Hainsworth and David Robey
DARWIN Jonathan Howard
THE DEAD SEA SCROLLS Timothy H. Lim
DECADENCE David Weir
DECOLONIZATION Dane Kennedy
DEMOCRACY Bernard Crick
DEMOGRAPHY Sarah Harper
DEPRESSION Jan Scott and Mary Jane Tacchi
DERRIDA Simon Glendinning
DESCARTES Tom Sorell
DESERTS Nick Middleton
DESIGN John Heskett
DEVELOPMENT Ian Goldin
DEVELOPMENTAL BIOLOGY Lewis Wolpert
THE DEVIL Darren Oldridge
DIASPORA Kevin Kenny
DICTIONARIES Lynda Mugglestone
DINOSAURS David Norman
DIPLOMACY Joseph M. Siracusa
DOCUMENTARY FILM Patricia Aufderheide
DREAMING J. Allan Hobson
DRUGS Les Iversen
DRUIDS Barry Cunliffe
DYNASTY Jeroen Duindam
DYSLEXIA Margaret J. Snowling
EARLY MUSIC Thomas Forrest Kelly
THE EARTH Martin Redfern
EARTH SYSTEM SCIENCE Tim Lenton
ECONOMICS Partha Dasgupta
EDUCATION Gary Thomas
EGYPTIAN MYTH Geraldine Pinch
EIGHTEENTH-CENTURY BRITAIN Paul Langford
THE ELEMENTS Philip Ball
EMOTION Dylan Evans
EMPIRE Stephen Howe
ENERGY SYSTEMS Nick Jenkins
ENGELS Terrell Carver
ENGINEERING David Blockley
THE ENGLISH LANGUAGE Simon Horobin
ENGLISH LITERATURE Jonathan Bate
THE ENLIGHTENMENT John Robertson
ENTREPRENEURSHIP Paul Westhead and Mike Wright
ENVIRONMENTAL ECONOMICS Stephen Smith

ENVIRONMENTAL ETHICS
Robin Attfield
ENVIRONMENTAL LAW
Elizabeth Fisher
ENVIRONMENTAL POLITICS
Andrew Dobson
EPICUREANISM Catherine Wilson
EPIDEMIOLOGY Rodolfo Saracci
ETHICS Simon Blackburn
ETHNOMUSICOLOGY Timothy Rice
THE ETRUSCANS Christopher Smith
EUGENICS Philippa Levine
THE EUROPEAN UNION
Simon Usherwood and John Pinder
EUROPEAN UNION LAW
Anthony Arnull
EVOLUTION Brian and
Deborah Charlesworth
EXISTENTIALISM Thomas Flynn
EXPLORATION Stewart A. Weaver
EXTINCTION Paul B. Wignall
THE EYE Michael Land
FAIRY TALE Marina Warner
FAMILY LAW Jonathan Herring
FASCISM Kevin Passmore
FASHION Rebecca Arnold
FEDERALISM Mark J. Rozel and
Clyde Wilcox
FEMINISM Margaret Walters
FILM Michael Wood
FILM MUSIC Kathryn Kalinak
FILM NOIR James Naremore
THE FIRST WORLD WAR
Michael Howard
FOLK MUSIC Mark Slobin
FOOD John Krebs
FORENSIC PSYCHOLOGY
David Canter
FORENSIC SCIENCE Jim Fraser
FORESTS Jaboury Ghazoul
FOSSILS Keith Thomson
FOUCAULT Gary Gutting
THE FOUNDING FATHERS
R. B. Bernstein
FRACTALS Kenneth Falconer
FREE SPEECH Nigel Warburton
FREE WILL Thomas Pink
FREEMASONRY Andreas Önnerfors
FRENCH LITERATURE John D. Lyons
THE FRENCH REVOLUTION
William Doyle
FREUD Anthony Storr
FUNDAMENTALISM Malise Ruthven
FUNGI Nicholas P. Money
THE FUTURE Jennifer M. Gidley
GALAXIES John Gribbin
GALILEO Stillman Drake
GAME THEORY Ken Binmore
GANDHI Bhikhu Parekh
GARDEN HISTORY Gordon Campbell
GENES Jonathan Slack
GENIUS Andrew Robinson
GENOMICS John Archibald
GEOFFREY CHAUCER David Wallace
GEOGRAPHY John Matthews and
David Herbert
GEOLOGY Jan Zalasiewicz
GEOPHYSICS William Lowrie
GEOPOLITICS Klaus Dodds
GERMAN LITERATURE Nicholas Boyle
GERMAN PHILOSOPHY
Andrew Bowie
GLACIATION David J. A. Evans
GLOBAL CATASTROPHES Bill McGuire
GLOBAL ECONOMIC HISTORY
Robert C. Allen
GLOBALIZATION Manfred Steger
GOD John Bowker
GOETHE Ritchie Robertson
THE GOTHIC Nick Groom
GOVERNANCE Mark Bevir
GRAVITY Timothy Clifton
THE GREAT DEPRESSION AND
THE NEW DEAL Eric Rauchway
HABERMAS James Gordon Finlayson
THE HABSBURG EMPIRE
Martyn Rady
HAPPINESS Daniel M. Haybron
THE HARLEM RENAISSANCE
Cheryl A. Wall
THE HEBREW BIBLE AS LITERATURE
Tod Linafelt
HEGEL Peter Singer
HEIDEGGER Michael Inwood
THE HELLENISTIC AGE
Peter Thonemann
HEREDITY John Waller
HERMENEUTICS Jens Zimmermann
HERODOTUS Jennifer T. Roberts
HIEROGLYPHS Penelope Wilson
HINDUISM Kim Knott
HISTORY John H. Arnold

THE HISTORY OF ASTRONOMY Michael Hoskin
THE HISTORY OF CHEMISTRY William H. Brock
THE HISTORY OF CHILDHOOD James Marten
THE HISTORY OF CINEMA Geoffrey Nowell-Smith
THE HISTORY OF LIFE Michael Benton
THE HISTORY OF MATHEMATICS Jacqueline Stedall
THE HISTORY OF MEDICINE William Bynum
THE HISTORY OF PHYSICS J. L. Heilbron
THE HISTORY OF TIME Leofranc Holford-Strevens
HIV AND AIDS Alan Whiteside
HOBBES Richard Tuck
HOLLYWOOD Peter Decherney
THE HOLY ROMAN EMPIRE Joachim Whaley
HOME Michael Allen Fox
HOMER Barbara Graziosi
HORMONES Martin Luck
HUMAN ANATOMY Leslie Klenerman
HUMAN EVOLUTION Bernard Wood
HUMAN RIGHTS Andrew Clapham
HUMANISM Stephen Law
HUME A. J. Ayer
HUMOUR Noël Carroll
THE ICE AGE Jamie Woodward
IDENTITY Florian Coulmas
IDEOLOGY Michael Freeden
THE IMMUNE SYSTEM Paul Klenerman
INDIAN CINEMA Ashish Rajadhyaksha
INDIAN PHILOSOPHY Sue Hamilton
THE INDUSTRIAL REVOLUTION Robert C. Allen
INFECTIOUS DISEASE Marta L. Wayne and Benjamin M. Bolker
INFINITY Ian Stewart
INFORMATION Luciano Floridi
INNOVATION Mark Dodgson and David Gann
INTELLECTUAL PROPERTY Siva Vaidhyanathan
INTELLIGENCE Ian J. Deary
INTERNATIONAL LAW Vaughan Lowe
INTERNATIONAL MIGRATION Khalid Koser
INTERNATIONAL RELATIONS Paul Wilkinson
INTERNATIONAL SECURITY Christopher S. Browning
IRAN Ali M. Ansari
ISLAM Malise Ruthven
ISLAMIC HISTORY Adam Silverstein
ISOTOPES Rob Ellam
ITALIAN LITERATURE Peter Hainsworth and David Robey
JESUS Richard Bauckham
JEWISH HISTORY David N. Myers
JOURNALISM Ian Hargreaves
JUDAISM Norman Solomon
JUNG Anthony Stevens
KABBALAH Joseph Dan
KAFKA Ritchie Robertson
KANT Roger Scruton
KEYNES Robert Skidelsky
KIERKEGAARD Patrick Gardiner
KNOWLEDGE Jennifer Nagel
THE KORAN Michael Cook
KOREA Michael J. Seth
LAKES Warwick F. Vincent
LANDSCAPE ARCHITECTURE Ian H. Thompson
LANDSCAPES AND GEOMORPHOLOGY Andrew Goudie and Heather Viles
LANGUAGES Stephen R. Anderson
LATE ANTIQUITY Gillian Clark
LAW Raymond Wacks
THE LAWS OF THERMODYNAMICS Peter Atkins
LEADERSHIP Keith Grint
LEARNING Mark Haselgrove
LEIBNIZ Maria Rosa Antognazza
LEO TOLSTOY Liza Knapp
LIBERALISM Michael Freeden
LIGHT Ian Walmsley
LINCOLN Allen C. Guelzo
LINGUISTICS Peter Matthews
LITERARY THEORY Jonathan Culler
LOCKE John Dunn
LOGIC Graham Priest
LOVE Ronald de Sousa

MACHIAVELLI Quentin Skinner
MADNESS Andrew Scull
MAGIC Owen Davies
MAGNA CARTA Nicholas Vincent
MAGNETISM Stephen Blundell
MALTHUS Donald Winch
MAMMALS T. S. Kemp
MANAGEMENT John Hendry
MAO Delia Davin
MARINE BIOLOGY Philip V. Mladenov
THE MARQUIS DE SADE John Phillips
MARTIN LUTHER Scott H. Hendrix
MARTYRDOM Jolyon Mitchell
MARX Peter Singer
MATERIALS Christopher Hall
MATHEMATICAL FINANCE Mark H. A. Davis
MATHEMATICS Timothy Gowers
MATTER Geoff Cottrell
THE MEANING OF LIFE Terry Eagleton
MEASUREMENT David Hand
MEDICAL ETHICS Michael Dunn and Tony Hope
MEDICAL LAW Charles Foster
MEDIEVAL BRITAIN John Gillingham and Ralph A. Griffiths
MEDIEVAL LITERATURE Elaine Treharne
MEDIEVAL PHILOSOPHY John Marenbon
MEMORY Jonathan K. Foster
METAPHYSICS Stephen Mumford
METHODISM William J. Abraham
THE MEXICAN REVOLUTION Alan Knight
MICHAEL FARADAY Frank A. J. L. James
MICROBIOLOGY Nicholas P. Money
MICROECONOMICS Avinash Dixit
MICROSCOPY Terence Allen
THE MIDDLE AGES Miri Rubin
MILITARY JUSTICE Eugene R. Fidell
MILITARY STRATEGY Antulio J. Echevarria II
MINERALS David Vaughan
MIRACLES Yujin Nagasawa
MODERN ARCHITECTURE Adam Sharr
MODERN ART David Cottington
MODERN CHINA Rana Mitter
MODERN DRAMA Kirsten E. Shepherd-Barr
MODERN FRANCE Vanessa R. Schwartz
MODERN INDIA Craig Jeffrey
MODERN IRELAND Senia Pašeta
MODERN ITALY Anna Cento Bull
MODERN JAPAN Christopher Goto-Jones
MODERN LATIN AMERICAN LITERATURE Roberto González Echevarría
MODERN WAR Richard English
MODERNISM Christopher Butler
MOLECULAR BIOLOGY Aysha Divan and Janice A. Royds
MOLECULES Philip Ball
MONASTICISM Stephen J. Davis
THE MONGOLS Morris Rossabi
MOONS David A. Rothery
MORMONISM Richard Lyman Bushman
MOUNTAINS Martin F. Price
MUHAMMAD Jonathan A. C. Brown
MULTICULTURALISM Ali Rattansi
MULTILINGUALISM John C. Maher
MUSIC Nicholas Cook
MYTH Robert A. Segal
NAPOLEON David Bell
THE NAPOLEONIC WARS Mike Rapport
NATIONALISM Steven Grosby
NATIVE AMERICAN LITERATURE Sean Teuton
NAVIGATION Jim Bennett
NAZI GERMANY Jane Caplan
NELSON MANDELA Elleke Boehmer
NEOLIBERALISM Manfred Steger and Ravi Roy
NETWORKS Guido Caldarelli and Michele Catanzaro
THE NEW TESTAMENT Luke Timothy Johnson
THE NEW TESTAMENT AS LITERATURE Kyle Keefer
NEWTON Robert Iliffe
NIELS BOHR J. L. Heilbron
NIETZSCHE Michael Tanner
NINETEENTH-CENTURY BRITAIN Christopher Harvie and H. C. G. Matthew

THE NORMAN CONQUEST
George Garnett
NORTH AMERICAN INDIANS
Theda Perdue and Michael D. Green
NORTHERN IRELAND
Marc Mulholland
NOTHING Frank Close
NUCLEAR PHYSICS Frank Close
NUCLEAR POWER Maxwell Irvine
NUCLEAR WEAPONS
Joseph M. Siracusa
NUMBERS Peter M. Higgins
NUTRITION David A. Bender
OBJECTIVITY Stephen Gaukroger
OCEANS Dorrik Stow
THE OLD TESTAMENT
Michael D. Coogan
THE ORCHESTRA D. Kern Holoman
ORGANIC CHEMISTRY
Graham Patrick
ORGANIZATIONS Mary Jo Hatch
ORGANIZED CRIME
Georgios A. Antonopoulos and
Georgios Papanicolaou
ORTHODOX CHRISTIANITY
A. Edward Siecienski
PAGANISM Owen Davies
PAIN Rob Boddice
THE PALESTINIAN-ISRAELI
CONFLICT Martin Bunton
PANDEMICS Christian W. McMillen
PARTICLE PHYSICS Frank Close
PAUL E. P. Sanders
PEACE Oliver P. Richmond
PENTECOSTALISM William K. Kay
PERCEPTION Brian Rogers
THE PERIODIC TABLE Eric R. Scerri
PHILOSOPHY Edward Craig
PHILOSOPHY IN THE ISLAMIC
WORLD Peter Adamson
PHILOSOPHY OF BIOLOGY
Samir Okasha
PHILOSOPHY OF LAW
Raymond Wacks
PHILOSOPHY OF SCIENCE
Samir Okasha
PHILOSOPHY OF RELIGION
Tim Bayne
PHOTOGRAPHY Steve Edwards
PHYSICAL CHEMISTRY Peter Atkins
PHYSICS Sidney Perkowitz
PILGRIMAGE Ian Reader
PLAGUE Paul Slack
PLANETS David A. Rothery
PLANTS Timothy Walker
PLATE TECTONICS Peter Molnar
PLATO Julia Annas
POETRY Bernard O'Donoghue
POLITICAL PHILOSOPHY
David Miller
POLITICS Kenneth Minogue
POPULISM Cas Mudde and
Cristóbal Rovira Kaltwasser
POSTCOLONIALISM Robert Young
POSTMODERNISM Christopher Butler
POSTSTRUCTURALISM
Catherine Belsey
POVERTY Philip N. Jefferson
PREHISTORY Chris Gosden
PRESOCRATIC PHILOSOPHY
Catherine Osborne
PRIVACY Raymond Wacks
PROBABILITY John Haigh
PROGRESSIVISM Walter Nugent
PROJECTS Andrew Davies
PROTESTANTISM Mark A. Noll
PSYCHIATRY Tom Burns
PSYCHOANALYSIS Daniel Pick
PSYCHOLOGY Gillian Butler and
Freda McManus
PSYCHOLOGY OF MUSIC
Elizabeth Hellmuth Margulis
PSYCHOPATHY Essi Viding
PSYCHOTHERAPY Tom Burns and
Eva Burns-Lundgren
PUBLIC ADMINISTRATION
Stella Z. Theodoulou and
Ravi K. Roy
PUBLIC HEALTH Virginia Berridge
PURITANISM Francis J. Bremer
THE QUAKERS Pink Dandelion
QUANTUM THEORY
John Polkinghorne
RACISM Ali Rattansi
RADIOACTIVITY Claudio Tuniz
RASTAFARI Ennis B. Edmonds
READING Belinda Jack
THE REAGAN REVOLUTION Gil Troy
REALITY Jan Westerhoff
THE REFORMATION Peter Marshall

RELATIVITY Russell Stannard
RELIGION IN AMERICA Timothy Beal
THE RENAISSANCE Jerry Brotton
RENAISSANCE ART Geraldine A. Johnson
REPTILES T. S. Kemp
REVOLUTIONS Jack A. Goldstone
RHETORIC Richard Toye
RISK Baruch Fischhoff and John Kadvany
RITUAL Barry Stephenson
RIVERS Nick Middleton
ROBOTICS Alan Winfield
ROCKS Jan Zalasiewicz
ROMAN BRITAIN Peter Salway
THE ROMAN EMPIRE Christopher Kelly
THE ROMAN REPUBLIC David M. Gwynn
ROMANTICISM Michael Ferber
ROUSSEAU Robert Wokler
RUSSELL A. C. Grayling
RUSSIAN HISTORY Geoffrey Hosking
RUSSIAN LITERATURE Catriona Kelly
THE RUSSIAN REVOLUTION S. A. Smith
SAINTS Simon Yarrow
SAVANNAS Peter A. Furley
SCEPTICISM Duncan Pritchard
SCHIZOPHRENIA Chris Frith and Eve Johnstone
SCHOPENHAUER Christopher Janaway
SCIENCE AND RELIGION Thomas Dixon
SCIENCE FICTION David Seed
THE SCIENTIFIC REVOLUTION Lawrence M. Principe
SCOTLAND Rab Houston
SECULARISM Andrew Copson
SEXUAL SELECTION Marlene Zuk and Leigh W. Simmons
SEXUALITY Véronique Mottier
SHAKESPEARE'S COMEDIES Bart van Es
SHAKESPEARE'S SONNETS AND POEMS Jonathan F. S. Post
SHAKESPEARE'S TRAGEDIES Stanley Wells
SIKHISM Eleanor Nesbitt
THE SILK ROAD James A. Millward
SLANG Jonathon Green
SLEEP Steven W. Lockley and Russell G. Foster
SOCIAL AND CULTURAL ANTHROPOLOGY John Monaghan and Peter Just
SOCIAL PSYCHOLOGY Richard J. Crisp
SOCIAL WORK Sally Holland and Jonathan Scourfield
SOCIALISM Michael Newman
SOCIOLINGUISTICS John Edwards
SOCIOLOGY Steve Bruce
SOCRATES C. C. W. Taylor
SOUND Mike Goldsmith
SOUTHEAST ASIA James R. Rush
THE SOVIET UNION Stephen Lovell
THE SPANISH CIVIL WAR Helen Graham
SPANISH LITERATURE Jo Labanyi
SPINOZA Roger Scruton
SPIRITUALITY Philip Sheldrake
SPORT Mike Cronin
STARS Andrew King
STATISTICS David J. Hand
STEM CELLS Jonathan Slack
STOICISM Brad Inwood
STRUCTURAL ENGINEERING David Blockley
STUART BRITAIN John Morrill
SUPERCONDUCTIVITY Stephen Blundell
SUPERSTITION Stuart Vyse
SYMMETRY Ian Stewart
SYNAESTHESIA Julia Simner
SYNTHETIC BIOLOGY Jamie A. Davies
TAXATION Stephen Smith
TEETH Peter S. Ungar
TELESCOPES Geoff Cottrell
TERRORISM Charles Townshend
THEATRE Marvin Carlson
THEOLOGY David F. Ford
THINKING AND REASONING Jonathan St B. T. Evans
THOMAS AQUINAS Fergus Kerr
THOUGHT Tim Bayne
TIBETAN BUDDHISM Matthew T. Kapstein
TIDES David George Bowers and Emyr Martyn Roberts
TOCQUEVILLE Harvey C. Mansfield

TOPOLOGY Richard Earl
TRAGEDY Adrian Poole
TRANSLATION Matthew Reynolds
THE TREATY OF VERSAILLES Michael S. Neiberg
TRIGONOMETRY Glen Van Brummelen
THE TROJAN WAR Eric H. Cline
TRUST Katherine Hawley
THE TUDORS John Guy
TWENTIETH-CENTURY BRITAIN Kenneth O. Morgan
TYPOGRAPHY Paul Luna
THE UNITED NATIONS Jussi M. Hanhimäki
UNIVERSITIES AND COLLEGES David Palfreyman and Paul Temple
THE U.S. CONGRESS Donald A. Ritchie
THE U.S. CONSTITUTION David J. Bodenhamer
THE U.S. SUPREME COURT Linda Greenhouse
UTILITARIANISM Katarzyna de Lazari-Radek and Peter Singer
UTOPIANISM Lyman Tower Sargent
VETERINARY SCIENCE James Yeates
THE VIKINGS Julian D. Richards
VIRUSES Dorothy H. Crawford
VOLTAIRE Nicholas Cronk
WAR AND TECHNOLOGY Alex Roland
WATER John Finney
WAVES Mike Goldsmith
WEATHER Storm Dunlop
THE WELFARE STATE David Garland
WILLIAM SHAKESPEARE Stanley Wells
WITCHCRAFT Malcolm Gaskill
WITTGENSTEIN A. C. Grayling
WORK Stephen Fineman
WORLD MUSIC Philip Bohlman
THE WORLD TRADE ORGANIZATION Amrita Narlikar
WORLD WAR II Gerhard L. Weinberg
WRITING AND SCRIPT Andrew Robinson
ZIONISM Michael Stanislawski

Available soon:

SMELL Matthew Cobb
THE SUN Philip Judge
AERIAL WARFARE Frank Ledwidge
RENEWABLE ENERGY Nick Jelley
NUMBER THEORY Robin Wilson

For more information visit our website

www.oup.com/vsi/

J. L. Heilbron

NIELS BOHR

A Very Short Introduction

OXFORD
UNIVERSITY PRESS

Great Clarendon Street, Oxford, OX2 6DP,
United Kingdom

Oxford University Press is a department of the University of Oxford.
It furthers the University's objective of excellence in research, scholarship,
and education by publishing worldwide. Oxford is a registered trade mark of
Oxford University Press in the UK and in certain other countries

First edition published in 2020
Impression: 1

Published in the United States of America by Oxford University Press
198 Madison Avenue, New York, NY 10016, United States of America

British Library Cataloguing in Publication Data
Data available

Library of Congress Control Number: 2019954874

ISBN 978–0–19–881926–4

Printed in Great Britain by
Ashford Colour Press Ltd, Gosport, Hampshire

Contents

List of illustrations

Prologue

Isaac Newton thought that the mutual gravitational pulls of the planets would cause them to fall into the sun unless God chose to top up their velocities. After a century of calculation, however, physicists decided that God had got it right after all, insofar as the stability of the solar system is concerned. Physics thus solved a problem (call it a God problem) that menaced its existence.

The nuclear atom returned physicists to the problem of the stability of their world. Electrons are not planets: they repel one another, and their mutual repulsion makes a nuclear atom with more than one electron mechanically unstable. Another damaging negative analogy exists between an electron and a planet. Physical theory before Bohr required that an accelerated charged particle radiate energy; consequently, even a one-electron nuclear atom would quickly collapse since the centripetal force needed to keep the electron in a circular orbit would cause it to radiate away its energy and dive into the nucleus. In his early work, Bohr called this menacing theory 'ordinary physics'. It would soon be dubbed 'classical physics' in the same sense that adventurous composers were beginning to call their heritage 'classical music'. Classical physics does not allow a nuclear atom whose parts obey its rules to exist.

And yet when Ernest Rutherford adopted the nuclear atom in 1911, he had good reason for supposing that it represented things as they are. In experiments done at the University of Manchester under his direction, matter behaved as if made of atoms with minute centres carrying electric charges much larger than the charge e on an electron. Calculation showed that this big charge, Ne, grew with the atomic weight A of the material investigated in a strikingly simple way: $N \approx A/2$. This close approximation had a parallel in measurements made by J. J. Thomson and his assistants at the Cavendish Laboratory in Cambridge, which indicated that the number n of electrons in an atom was also around $A/2$. Rutherford's minute centre could then be understood as a 'nucleus' bearing a charge equal and opposite to the sum of the charges N on the planetary electrons: $N \approx n$.

A consequence of utmost importance followed from this simple equation. In Rutherford's experiments, alpha particles from radioactive substances bombarded metal targets. In calculating the outcome of collisions between the particles and the nucleus he assumed that he could neglect their sizes: that is, he considered both parties as bare nuclei. Since earlier he had demonstrated that an alpha particle consisted of a helium atom minus two electrons, it followed that the neutral helium atom has precisely two planetary electrons. A reasonable inference gave hydrogen one electron, lithium three electrons, beryllium four, and so on throughout the periodic table of elements. Thus it was natural to conclude that Z, the ordinal or 'atomic number' of an element in the periodic table, equalled the number of charges on the nucleus N. Of equal importance, Rutherford's atom allowed a rigid distinction between phenomena arising from the electronic planetary system, which the experimenter could influence, and radioactivity, seated in the nucleus, and apparently immune from external forces. A pity that Rutherford's model was incurably unstable!

Niels Bohr was an advanced research student at Manchester in the spring and summer of 1912. Although he had come to England in the autumn of 1911 intending to work with Thomson on the properties of metals, the subject of his Danish doctoral thesis, the collaboration did not materialize, and Bohr decided to finish his fellowship year learning something about radioactivity from Rutherford. He soon took an interest in his new professor's new atom.

The interest became a passion when Bohr discovered for himself that ordinary physics did not permit nuclear atoms. Like Newton, he recognized the need for some new principle that limited or corrected the application of ordinary physics and thought himself capable of finding it. He could not, of course, appeal to God, in whom in any case he did not believe; but he would need to lay down a principle with God-like authority in order to enable nature to make, and physicists to stabilize, atoms with orbiting electrons. It would require the insight of an Einstein and the confidence of a Viking; for Bohr's refashioning of physics would demand greater sacrifices than the relaxation of our intuitions of space and time claimed by the theory of relativity.

Bohr spent much of his life challenging his own thinking, writing and rewriting, trying to find the limits of applicability of descriptions that seemed contradictory. He came to identify his great task as teaching physicists, and then the rest of humankind, how to express themselves without ambiguity while using, as necessary, concepts ordinarily regarded as contradictory. No wonder he had trouble saying what he meant! He communicated the art of communicating unambiguously ambiguously, sometimes oracularly. 'The opposite of a deep truth is a deep truth,' he would say, and 'the opposite of a mere truth is clarity'. Like any good oracle he sometimes communicated in parables. A favourite concerned a bright boy reporting on the lectures of a profound rabbi. The first lecture was clear and precise: it concerned mere truths, and the boy understood every word. Of the second he understood little, but the rabbi grasped it all. The third, the boy said, was sublime. It dealt with deep truths; he understood not a word and the rabbi not much more.

Bohr had worried about truth and its unambiguous communication long before he came to the nuclear atom. He took these problems from the teachings of his professor of philosophy at the University of Copenhagen, Harald Høffding, and the writings of Søren Kierkegaard, which he read assiduously while studying for his master's degree in physics. The problems of old Danish philosophy and the difficulties of the latest physics were the equipment with which Bohr's bold mind met the challenge of the nuclear atom. He had two strong motivations for taking up a God problem. One was an intense adolescent wrestle with religion. The other was a characteristic scruple: Bohr felt a need to demonstrate to family and friends that they were not mistaken in encouraging him to devote his life to thinking.

One of the problems that guided his early thinking was the opposition between free will and determinism. Accepting that we need both concepts to describe our experience, he worked out how to use them without contradiction: free will refers to the contemplation of future acts, determinism to the analysis of completed acts. They have to do with different situations. Bohr returned to this example of what Einstein later denigrated as a 'tranquilizing philosophy' after the development of his non-classical nuclear atom had sharpened the tools he needed to extend his insights about multiple truths and unambiguous language to realms of thought and action beyond physics.

Bohr's engagement with physics and its lessons went beyond the theoretical. He proved to be an effective fundraiser for the Institute set up for him in Copenhagen in 1922 and guided it into research in nuclear physics and radiobiology. During the Second World War he worked briefly at Los Alamos and more perseveringly but less effectively at convincing Roosevelt and Churchill to share information about atomic weapons with Stalin in order to avoid a post-war arms race. Some of the rhetoric he employed in this endeavour came straight from the lessons he derived from physics. Just as physicists had to 'renounce' causal descriptions in space and time in order to control their descriptions of the atomic world, so politicians would have to renounce old-fashioned sovereignty in order to control the weaponry atomic physicists had given them.

Bohr believed that orderly and apt renunciation of concepts or goals previously thought unchallengeable was the key to freeing our minds from pseudo-problems and superstitions inferred from ordinary experience. We advance by learning what to give up. That is why Galileo was so important; he managed to 'liberate himself [from misleading sensory impressions] and renounce any explanation of motion'. Bohr preached the doctrine of renunciation in physics, sometimes to great success (as in his limitation of ordinary physical concepts in the atomic domain), and sometimes to no avail (as in his occasional proposals to abandon the conservation of energy). His urging of the surrender of a degree of national sovereignty in the interest of world stability, notably in an open letter to the United Nations in 1950, fared no better than his proposed amendment of the first law of thermodynamics. Nor did he succeed in establishing 'complementarity'—his great lesson from quantum physics—as an antidote to religion and a guide to the improvement of humanity.

For several reasons—the obscurity of his writing and dislike of mathematical formalism, the failure of complementarity to remain alive outside philosophy, and the relegation of his atomic model to the care of historians—Bohr does not enjoy the same recognition, even among physicists, as his friend and intellectual equal Einstein. One purpose of this little book is to present the debt physics owes to Bohr's leadership and to make plausible the view expressed by the President of the Royal Society of London in 1944, that 'a vote of the world's scientists would place [Bohr] first among all the men of all countries who are now active in any department of science'. Another objective, perhaps more useful and realizable, is to illustrate a powerful thinker's striving to reach to the bottom of things. That striving is what Bohr admired in

Kierkegaard and lines from Goethe's *Weite Welt und breites Leben*. In English they read:

> Tireless searching, firmly founded
> Never ended, often rounded
> Old traditions well respected
> Innovations not rejected.

In pursuit of these objectives some ideas and concepts from physics will be needed. Uncommon ones will be defined as they come up. Some mathematical symbols appear from time to time. They are shorthand employed to avoid repetition and because Bohr could not do without them, even though he did not always know exactly what they meant. They too will be defined as needed. There will be no appeal to mathematics beyond the slightest algebra or to philosophical technicalities. Everything will gradually become clear to anyone willing to acquire the minimal equipment for accompanying Bohr, who did not know where he was going either, on his extraordinary intellectual journey.

Chapter 1
A richly furnished mind

'Viking Jew'

The very original mind of Niels Bohr came into existence in Copenhagen in 1885. It arose from the union of a professor of physiology at the city's university, Christian Bohr, and Ellen Adler, the daughter of a liberal philanthropic Jewish banker (Figures 1 and 2). Although (and perhaps because) Christian came from a family of pastors, he was an atheist; and although Ellen's family were prominent among 'Viking Jews'—Jews who settled and prospered in Denmark before the immigration of Yiddish speakers from the shtetl—she was not observant. Nonetheless, some accommodation for the religious training of their three children was required. Christian agreed that they be raised as Jews. Ellen did not insist, however, and Niels and his brother Harald were sent to the Lutheran church so that they would have the same religious experience as other Danish boys. Of the couple's eldest child, Jenny, few accessible records exist.

The forces of parental indifference and state religion had to do battle with Ellen's sister Hanna, the first woman to earn an advanced degree in physics in Denmark and the proprietor of the first coeducational secondary school in the kingdom. Aunt Hanna (Figure 3) was close to her nephews, especially Niels. An expert

1. Bohr's father, Christian Bohr.

on the history of Danish Jewry cites Hanna Adler's work as exemplary of the 'special, very concrete idealism, expressed in an intensive and persistent striving to realize an idea or thought', that constituted the Jewish 'impulse' in Danish science. In contrast

2. Bohr's mother, Ellen Adler.

with her sister Ellen, who had little or no connection with the synagogue, Hanna had close ties to the Jewish community.

Naturally religion loomed large in Niels's adolescent mind as it struggled to balance the views of his parents, his pastors, and his aunt. One day he told his father that although he had struggled very, very hard, he could not believe. As he later described the incident to his fiancée, Margrethe Nørlund, his father responded with a smile. Bohr interpreted the smile as a sign that 'I too could think.' For a time he thought to write a book to help others think

3. Aunt Hanna keeping in touch with her nephews.

themselves free from the errors of organized religion, and echoes of this unrealized project may be discerned in other forms in his later writings.

Unbelievers are of course not free from the cultures in which they grow up. In Bohr's case both Christian and Jewish elements continued to occupy his mind until the simultaneous arrival of his wedding and his quantized atom filled it with other things. The Jewish element of his boyhood came not only from Aunt Hanna, but also from his Adler cousins, notably Edgar Rubin, who became a leading experimental psychologist. Rubin had an opportunity to observe some extraordinary minds in a club of university students dominated by himself and the Bohr brothers. Many of the topics debated in the Ekliptika, as the twelve overachievers who constituted the club called it, started from Høffding's teaching. At least half of this academic zodiac were Jewish.

The remarkable representation of students of Jewish descent in the Ekliptika and their subsequent high academic achievement were consonant with the traditional Jewish emphasis on study, the tendency of liberal reform Jews to assimilate, and the relative tolerance of Danish society. Five of the members were cousins of one sort or another, in keeping with the tendency, even among assimilated Jews, to stick together. Two of the gentile members of the club, the brothers Niels Erik Nørlund (a mathematician) and Poul Nørlund (a historian), joined this large family when their younger sister Margrethe became engaged to Niels (Figure 4).

When Bohr returned to Copenhagen from Manchester during the Easter vacation of 1912 to help plan his wedding, the cultural hit of the season was an improbably popular play, *Indenfor murene* ('Inside the walls'), by the Danish Zionist Henri Nathansen. It tells the story of a Jewess, Esther, who falls in love with a gentile professor whose lectures she attended, just as Ellen Adler had done with Christian Bohr. Esther's engagement distresses both families, but the action takes place primarily 'within the walls', in the warm, cosy, middle-class Jewish home that Esther's desertion threatens to destroy (Figure 5). Similarly, the story of Niels and Margrethe unfolded within the Adler side of the family; Margrethe lived with Ellen Bohr for some time before her marriage and saw very little of the Bohr side and its Lutheran theologians. But whereas Margrethe joined the assimilated Adlers easily, Esther could not be comfortable with her bigoted in-laws.

The real-life story of another member of the Ekliptica, Lis Rubin Jacobsen, who received her doctorate in 1910 when already married and became an authority on Nordic runes, is similar and similarly cogent. Her non-observant father Marcus Rubin was close to the leading Danish-Jewish man of letters, Georg Brandes, who fervently favoured assimilation; yet Marcus would not allow

4. Niels Bohr and Margrethe Nørlund at their engagement.

his daughter to attend Christian religious instruction in school. When she asked permission to participate in these prohibited exercises, Marcus gave her some Christian books to read. That was enough. She grew up an unbeliever and aimed to be a schoolteacher until she fell in love with a gentile intellectual.

5. 'Inside the walls': the home of Bohr's Jewish grandparents.

Though without faith, married into a Christian family, and urged to assimilation by Brandes, she could not shake off her Jewish identity. In later life she promoted Jewish causes and is now cited along with Hanna Adler among the most distinguished Jewish women in Danish history.

Bohr and Jacobsen shared the traits to which Nathansen ascribed Jewish success in surviving persecution. The leading items in his inventory, as itemized in Nathansen's sensitive biography of his friend Brandes, are strength and joy in work. To the world Jews are fierce competitors, who can come across as over-critical, domineering, and arrogant. Yet their struggle for equality made them champions of truth, justice, freedom, and human rights. With family and friends, 'inside the walls', Nathansen's Jews have an 'intimate special life…whose passwords [are] "respect" and "discipline"—respect for tradition, discipline in the family'. There competitiveness turns to humour, irony, satire, word play, banter, 'wily, equivocal, ambiguous, double-edge wit combined with irony and self-irony'. 'The world of the mind was the home of the homeless Jewish people, the life of the mind their only free state.... From the special exclusivity of this life of the mind Jewish "chutzpa", boldness, something between courage and insolence, has developed, and also the Jewish "chain", the artistic, sensitive union of grace and taste, something between enchantment and enticement.' Niels Bohr refined and combined these characteristics into his special ways of thought and expression.

While planning his wedding Bohr encountered a religious problem of the same intensity as the difficulty that had precipitated his loss of faith. He refused to be married in church, from which he and Margrethe resigned; the decision upset her pious mother, whose hurt Bohr tried to assuage with the reassurance that he did not believe that science governed or could govern everything. Moreover, he thought that he could prove it. And that, he said, was a source of joy to him: '[life] would be so infinitely trivial if I thought I could understand it'. Thus he was particularly aware of the necessity and difficulty of reconciling conflicting religious beliefs and cultures, and of the probable existence of aspects of experience beyond rational explanation, just before he encountered the contradictions of the nuclear atom.

The Viking in Bohr's make-up was not only Jewish. He had a great love for the North, the land of volcanoes and glaciers, fire and ice, mirages, glories, and auroras—and of sagas, heroes, giants, kobolds, trolls, snow queens, and ice palaces. There was romance in his make-up. 'When I see the briefest reference to the old Nordic countries, then my heart flares up so wildly, so wildly, my little one, I dream I am among Norway's cliffs and skerries.' The occasion of this outburst was reading the chapter on Odin in Thomas Carlyle's *On heroes and hero worship*, which the little one, Margrethe, had sent him as an example of her favourite reading. He went on to pose a pair of crazy questions: '[T]ell me whether you would come with me to Iceland in a Viking ship; tell me whether you would, and tell me, whether you, in addition, would stay behind on your own in Iceland, when I had to leave in the summer.' She replied in the literary idiom in which they clothed their emotions. 'I will come to you, Niels, as Solveig came to Peer Gynt.' This was to grant more than was required, however, since the innocent Solveig, who offered herself freely to the vagabond Peer, wasted her life waiting for his return. The catechism continued. 'Will you care for my work?' 'Dear Niels, I cannot at all describe to you how much I love you and how much I love your work.' But will you be a mother to my students? 'I set no limit at all to how much I wish that I could be allowed to be a mother to your students.' The limit would be the size of Bohr's Institute.

Another indication of the content of Bohr's mind around the time of his great invention is a story that he wrote for Margrethe, then still his fiancée, for Christmas of 1911. The story begins with their watching a little boy accompanying his father to church on Christmas day, the only day in the year the father entered a church; the little boy soon discovered, 'after his own lonely battles', the reason that his father otherwise kept away. '[The boy] could not imagine anything as dreadful and terrible' as the realization of the things taught in church. After doing honour to this perfect parent, then recently deceased, the couple fly to the far north, where they again spy the boy, now an adult, discussing philosophy with an old gentleman who no doubt answered to the name of Høffding. Eventually they return to Bohr's digs in Cambridge, whence Margrethe modestly leaves for Denmark while Niels retires to bed, 'his courage roar[ing] so wildly, so wildly, for he thinks that he too could think'. With Margrethe's help, he will try to combine courage and thinking, Viking virtue with something else: 'My own little darling, if you will care for him, he will try to find meaning in his wild courage.'

Then came a question unusual for a Viking. Will you pay my debts, 'all the debts that my poor soul might incur?' Although Niels repeated this question several ways, he never specified the obligations he needed Margrethe's help to discharge. Perhaps he meant that with her psychological support he could justify, and so repay, the belief in his abilities entertained by his family, teachers, and friends. To his mother he was a 'rare treasure', to his father 'gold', to his brother, 'the greatest and wisest human being we have known'. The family had helped him to develop the gifts of nature. He dictated most of his thesis to his mother. His father put his laboratory and assistant at Niels's disposal when he needed them. The entire family helped in computing tables, doing calculations, and writing out fair copies of his papers. Margrethe as amanuensis and also her brother Niels Erik as calculator became part of this machinery even before her marriage. There was much to repay.

The Viking in Bohr may also be seen in his strength and endurance as a skier and football player. He could ski for days and his size made him a good goalie as long as he kept his mind on the game. Harald was more agile in sport as well as more practical in life than Niels and made it to the Danish Olympic football team.

Christian philosopher

Even before entering the university, Bohr spent time with Høffding (Figure 6). The philosopher was a great friend of Christian Bohr and also of Christian Christiansen, who would be Bohr's professor of physics. The three met regularly with the philologist Vilhelm Thomsen, and Niels and Harald would listen to them airing deep questions of science and philosophy when Christian hosted their symposium. This eavesdropping gave the brothers at least two precious insights: that experienced scholars can advance their discipline before securing its foundations and that responsible scholars strive to deepen as well as enlarge their discipline's structure. The most basic of the foundational issues the quartet of professors discussed concerned 'the nature, condition and limits of knowledge, the nature and worth of evidence, and the principles which underlie our valuation of human actions and institutions', that is, the problem of Truth.

6. Bohr's professor of philosophy, Harald Høffding.

Høffding had a lot to say about truth. He began with Søren Kierkegaard, whom Bohr was to rank as the greatest of Danish thinkers and stylists, and the author of one of the best books ever written. This was *Stages on life's way*, which Høffding took to be representative of Kierkegaard's philosophy and a great help to people undergoing religious crises. It had helped Høffding himself in early life, and his writing about truth, like Kierkegaard's, owed much to struggling with a Christian worldview. He arrived at the capacious principle that no single truth can capture a domain, for no matter how promising a line of analysis is at the beginning, if pushed ever further it must eventually expose an irremediable, inaccessible, irrational residuum. This was the proposition that Bohr told Margrethe's mother he could prove logically. Enthusiasm over the necessity of renouncing the search for a theory of everything marked Høffding's modest epistemology; as Rubin recalled, 'this state of affairs caused him great and profound satisfaction', for, like Bohr, he regarded its contrary, in which everything would stand revealed, as the destruction of 'an essential condition for the value of human life'.

Høffding had begun his university studies in theology. After a long internal fight guided by Kierkegaard's similar struggle he decided that he could not live his life 'by the ideals and commandments of religious ethics' and looked to philosophy to find 'equivalents for the loss of belief in those goods which the vanishing of religion entails'. Bohr gave expression to the same programme when he assured his future mother-in-law that he believed in many things: 'in the goodness and love of human beings, for that I have experienced'; 'in the duties of a human being, although I cannot say exactly what they are'; and 'in so many many other things that I do not understand'. How can these things be justified, grounded, in absence of religion? Bohr could only hope, 'with all my soul' and without supernatural help or threats, that he could stay true to his ideals of 'the good and great and true'. This was a moral, if not a philosophical solution, to the great problem of Truth, which, as he knew from Høffding, could not be solved.

For Høffding, free enquiry in the religious sphere was the pre-eminent means for awakening and encouraging thought. 'He to whom the problem [of religion] does not present itself has of course no ground for thought, but neither has he any ground for preventing other people from thinking.' Høffding's even-handed consideration of religion persuaded his students, 'for whom his lectures were the experience of their university years', and worried their parents, who feared, rightly, that he might dissolve their traditional beliefs. He put an extravagant value on intellectual life. So did young Bohr ('it is the most valuable and only thing I possess'), who hoped to enter the only class of scientist that, according to Høffding, required true scientific culture. These were the creators of new theories.

Høffding stayed in contact with the Bohr family after Christian Bohr's death in 1911. In his old age his 'good friend Niels Bohr' would visit him to talk about physics and philosophy, and read from their favourite poets, for 'Niels Bohr is not only a great physicist, but also is interested in philosophy and literature.' Høffding was able to make use of Bohr's ideas in a widely published essay on the concept of analogy and Bohr, returning the compliment, credited Høffding with 'ideas that helped physicists to "understand" their work'. On Høffding's death, Bohr succeeded him in the *Aeresbolig*, the villa left by the founder of the Carlsberg Brewery as the home of the greatest intellectual among the Danes as determined by the Danish Academy of Sciences. The succession might serve as a symbol of Bohr's place in Danish philosophy and culture.

Viking philosopher

Only the boldest of thinkers could expect to refashion his science with the very same tools he used to prove 'logically' that human beings can never know everything. And how is that proved? Let us begin with physics. Our mode of understanding is continuity of thought; hence physics posits continuity in motion, in action, in cause and effect; as the old philosophers said, nature does not make jumps. Is this assumption a requirement of thought? 'The great question is, whether the idea of the continuity of motion or activity can be carried out in all spheres.' If not, room opens for 'an irrational relationship' between nature and knowledge. And that in fact is how things are. 'For us, existence can never be absorbed into thought without remainder.'

Our usual analyses suppose a clean division between the subject (the observer) and the object (the observed). Alas, it is but an indulgent delusion. Object and subject mutually determine one another: a pure subject is as illusory as a thing-in-itself. Not only is there no pure case, but no place to stop: the subject influenced by the object becomes a new observer of a new object, and so on and on. 'Here again we run up against the irrational and here perhaps we see most clearly how inexhaustible being is in comparison to our knowledge.' There is no reason for despair in the realization that human beings cannot create 'an exhaustive concept of reality'; for it is just in 'the irrationality in the relation between thought and reality [that]... the possibility of progress lies'. Thus Høffding.

Kierkegaard says the same things even better. There is no way for us to create a complete account of Being because our knowledge and experience grow and change; and as we are part of the Being we are trying to capture in thought, we are attempting to grasp something unformed or continually forming. (Kierkegaard snickered that academic philosophers had missed this point because they are such non-entities they excluded themselves from existence in general.) This problem of the Subject altered by the Object can be traced back further, to a story by Kierkegaard's major patron, Poul Møller, a professor of philosophy considered by many to have been the archetypical Danish writer of his time. In Møller's story a student addicted to thought drives himself into intellectual impotence by thinking about himself thinking about a second self thinking... and into physical impotence by finding no sufficient reason to perform an action at any particular time and, hence, at no time at all. Bohr thought this story so expressive of the problems of quantum physics and the Danish way of handling them that he later urged it on all his foreign students as soon as they knew enough of the language to read it. For it presented not only the problem of the division between subject and object, but also the need to break off a logical line of thought arbitrarily in order to progress at all.

Introducing considerations of time provides another way of demonstrating limitations in the reach of rational analysis: whatever understanding we achieve can only be retrospective. As Høffding put the point in 1904, 'we live forward but understand backward'. Not everything lends itself to backwards comprehension: we will never be able to explain how we can understand retrospectively the necessity of what was open-ended prospectively. This was to phrase the problem of free will in precisely the terms in which Bohr approached it, by the doctrine of multiple partial truths: we are free in prospect, bound in retrospect. '[A] situation that calls for a description of our feeling of volition and a situation demanding that we ponder on the motives for our actions have quite different conscious contents.'

The irrationality of development in time appears arrestingly in Kierkegaard's *Stages on life's way*. Bohr had a copy of it with him at the rural parsonage to which he withdrew from the bustle of quiet Copenhagen to prepare for his master's thesis and examination. It was just the place for romantic intellectuals like Bohr and Kierkegaard. 'I walk here in solitude and think about so many things.' Thus Bohr. He thought about physics, of course, and mathematics and logic, but also about the problem of cognition, the stages of life, the nature of the good. The experience meant much to him as he could still relate it in accurate detail many years later. '[Kierkegaard] made a powerful impression on me when I wrote my dissertation at a parsonage on Funen, and I read his works day and night... His honesty and willingness to think the problems through to their very limit is what is great. And his language is wonderful, often sublime.' Bohr sent his copy of *Stages* to Harald as a birthday present from Funen. 'It is the only thing I have to send; nevertheless, I don't think I could easily find anything better. . . . I think absolutely that it is about the most beautiful thing that I have ever read.'

Kierkegaard's *Stages* employs six personae to convey his insights into the human condition. The earliest stage, the aesthetic, which for some people lasts a lifetime, is a period of carefree experimentation. Kierkegaard depicts it through speeches given by four of his avatars at a symposium on love, life, and the universe. Each says something true, though his statement conflicts with what the others say. Another avatar, a self-satisfied judge, sets forth the merits of a good marriage, the essence of the second or ethical stage. The judge's wife was patient, understanding, supportive, protective; neither he nor she could achieve as much apart as they did by pooling their complementary qualities; each contributed an equal share to the truths of married life. Bohr needed such a partner more than most men. Margrethe fitted the pattern perfectly. The third and final stage, the religious, can be reached only by a leap of faith, a quantum jump that, as we know, Bohr could not make.

Another of Kierkegaard's personae made a perfect model for a romantic young critic walking in solitude around a country personage. This was Johannes Climacus, who had a passion for thinking so intense that he could not think about girls. 'In love he was, madly in love, but with thought, or rather with thinking.' Having a 'romantic soul which always looked for difficulties', that is, being a consummate critic like Bohr, Climacus managed to prove that the foundational principle, 'modern philosophy begins with doubt', to which philosophers since Descartes had ascribed some meaning, meant nothing at all. Poor Climacus never advanced even to the threshold of received philosophy. 'He became more and more retiring, fearing that thinkers of distinction might smile at him when they heard that he too wanted to think.'

Physicist

The master's thesis Bohr was writing when he consulted Kierkegaard daily reviewed the electron theory of metals, which proposed to explain their properties of radiation, magnetism, and conduction on the implausible assumption that electrons flow through a wire as gas molecules move down a pipe. The assumption had enabled physicists to transfer the elaborate statistical methods they used with gases to the situation in a conducting wire. Bohr continued with his critical review of the electron theory in his doctoral dissertation, which he defended without opposition in 1911, as no one in Denmark had enough physics to judge it. Among other valuable results Bohr proved that the theory could not give an account of the radiation of heat from metals or their behaviour under a magnet in agreement with experiment.

The failure to account for radiation did not surprise him. A decade earlier a simpler case of heat radiation had resisted an aggressive attack by Max Planck, the leading theoretical physicist in Germany. Planck's simpler but more recondite problem was to determine the energy spectrum of the radiation in a hot oven whose walls are maintained at a constant temperature. By 'energy spectrum' he understood the relative intensities of the radiation at different frequencies, so much in the red region, so much in the yellow, so much in the blue, and so on. The goal was to specify a number ρ (the Greek letter 'rho') for the intensity of every detectable frequency ν (Greek 'nu') at the constant temperature T of the oven. Or, as physicists wrote the object of his quest, Planck sought a mathematical expression, $\rho(\nu, T)$, that summarized the state of the radiation.

Planck's oven problem attracted him because ordinary physics could prove that $\rho(\nu, T)$ did not depend on the material of the oven walls and so must have significance well beyond a description of the radiation in a particular heated cavity. Besides the quantity defining equilibrium (T) and the variable designating colour (ν), ρ could contain only constants characteristic of the two great departments of physics at whose intersection Planck's problem stood. These departments dealt with 'aether' (the medium in which light, heat radiation, and electromagnetic forces had their seat) and 'matter' (ponderable atoms in the various species itemized in the periodic table of elements). Up-to-date theorists in 1900 recognized the then recently detected electron, supposed to consist of a minute mass inseparably bound to an electric charge, as the likely coupler between aether and matter. In an extravagant extrapolation of this idea, J. J. Thomson had proposed to make the electron the building block of atoms, the key to the peculiar properties of metals and the periodic character of the elements, and, when in motion, the cause of most processes in the aether including radiation.

And how does an electron stir up the aether? This was one of the two 'clouds', or dark problems, that the dean of British mathematical physicists, Lord Kelvin, saw looming over physics at the start of the 20th century. The pith of the problem lay in the difficulty, widely acknowledged by 1900 as the impossibility, of describing the behaviour of aether with the concepts developed for ponderable matter. Kelvin pointed to experiments that had failed to detect any of the effects that should have arisen if aether had the least inclination to react mechanically to the motion of a ponderable body through it. Einstein dispelled this cloud in 1905 but, as Kelvin had foreseen, at a high cost; for the theory of relativity demands the surrender of intuitions of space and time. Kelvin was no less prescient in identifying his second cloud. It washed out Planck's attempt to subjugate cavity radiation to ordinary physics.

This second cloud was a democratic principle known as 'equipartition of energy'. An apparently inescapable conclusion of the representation of a gas sample as a huge assemblage of bouncing molecules, the principle required that at equilibrium and over time every molecule should enjoy the same average energy proportional to the equilibrium temperature. Equipartition proved catastrophic (that is how physicists described it) when applied to the aether in Planck's oven because it drove all the radiant energy into high frequencies. Planck's physics made his oven glow even when cold.

The reason that the upper modes in cavity radiation should batten on energy at the expense of the lower even with the equal suffrage of equipartition is easy to see. When equilibrium sets up in the cavity, the aether may be compared, despite its refusal to obey the laws of mechanics, to a gigantic collection of guitar strings each of which has its ends fixed to a wall. The fixing allows only vibrations whose wavelengths are integral subdivisions of the strings' lengths. The shorter the wavelength (and therefore the higher the frequency) the larger the number of waves that can fit along the string. Since equipartition assigns the same average energy to every mode, the upper end of the spectrum has almost all the energy when the radiation comes to equilibrium. To avoid this theoretical catastrophe, some of Thomson's colleagues, notably James Jeans, had the ingenious idea that the world has not existed long enough for equilibrium to take hold. Theirs was physics on God's scale.

Planck reversed the prejudice in favour of higher modes by arbitrarily placing a threshold on the energy needed to activate them. In effect he required the 'resonators' that made up the walls of the cavity to have a minimum activation energy ε proportional to their frequency, whence the expression $\varepsilon = h\nu$, which would become as famous in physics as $E = mc^2$. (A resonator is an electron fixed to a perfect weightless infinitesimal spring, available in all good physics laboratories, which emits precisely homogeneous radiation when vibrating.) The finite (though very small) value of 'Planck's constant' h makes resonators at higher frequencies struggle harder than those at lower frequencies to obtain their minimum 'quantum' $h\nu$. Its social equivalent would be to make it harder for rich people than for poor people to make money—an apt measure of the oddity of Planck's quantum theory.

Planck's resonators could take up and give off energy only in jumps and lumps $h\nu$. That violated concepts of continuity on which ordinary physics relied. And it left the sub-puzzles whether a resonator that accumulates more than one quantum can dispose of it all at once, or only in single quanta one at a time; and whether the emitted quantum causes waves in the aether as a rock does by falling into a pond, or sails through the pond as if it were not there. In 1905 Einstein gave reasons for believing that in some instances radiant energy traverses the aether like projectiles shot through a vacuum. (Einstein did not banish the medium to which physicists of his time gave the title 'aether': the 'space' or 'vacuum' of modern physics performs even more services.) Bohr rejected Einstein's 'light quanta' for some twenty years before experiment, forcing him to admit them, prompted the creation of his definitive mixture of philosophy and physics.

Fortunately, Bohr did not need to know the nature of light to finish his doctoral thesis of 1911. He worried rather about equipartition, which, as he discovered, destroyed a major success of the electron theory of metals. This was the ascription of para- and diamagnetism to the orientation of electron orbits in a magnetic field. Bohr demonstrated that equipartition did not allow the orientation to persist if the electrons interacted. To save the phenomena a mechanism foreign to the theory, something, Bohr hinted, analogous to Planck's restriction on resonators, had to be invoked to prevent promiscuous sharing of energy. Since Bohr expected all theories to break down somewhere, he no doubt was delighted to discover so clear a case of failure on his own. The problem with magnetism was one of the serious flaws in the theory of metals that Bohr planned to discuss with Thomson during the postdoctoral year he arranged to spend at Cambridge.

Bohr arrived in Cambridge in the autumn of 1911, detained, after sustaining his thesis in the spring, by the sudden death of his father in February and the revision of his thesis for the press. He had also to collect some money for a year's study abroad, which he did by a request, amounting in total to some twenty words, to the Carlsberg Foundation, and to secure, in many more words, an English translation of his thesis that he hoped to print in England.

Chapter 2
Productive ambiguity

Aether and matter

The Cambridge lions received Bohr cordially. Thomson arranged a college connection for him, and the professor of mathematical physics, Joseph Larmor, offered to publish his thesis if cut to half its length and expressed in comprehensible English. During his first few weeks he was ecstatic. 'I found myself rejoicing this morning,' he wrote to Margrethe on 26 September, 'when I stood outside a shop and by chance happened to read the address "Cambridge" over the door'. But the collaboration he had envisaged with Thomson did not work out. The professor had moved away from the theory of metals and did not have the time to master the Danish English of Bohr's thesis. Bohr worked diligently at translation and at the pedestrian experimental investigation Thomson gave him, but he had neither the interest nor the language to prosper in the laboratory. Instead he read and attended lectures. Among the lectures he praised was one by Thomson—interesting, instructive, beautiful, sparkling, scintillating—on the flight of a golf ball; for Bohr was a connoisseur ('a little crazy about') the ordinary physics he would reject in the atomic domain. Among the books that inspired him was Larmor's attempt at a theory of everything, *Aether and matter* (1900).

Larmor's difficult book taught the nobility of striving for a universal theory in the face of certain failure and the legitimacy of making assumptions, even unrealistic ones, if pursued honestly and consistently. Do not fear 'leaving reality behind... every result of thought [may be so] described which is more than a record or comparison of sensations'. Larmor had discovered that ordinary physics would not hold in the microworld. How then to proceed? Drop your scruples, raise your game, think, think, think; 'the history of discovery may be held perhaps to supply the strongest reason for estimating effort towards clearness of thought as of not less importance than exploration of phenomena'. There was encouragement for an ambitious student! 'When I read something that is so good and grand, then I feel such courage and desire to try whether I too could accomplish a tiny bit.'

Bohr soon decided that Cambridge was not the place to do his bit. The college system did not have a proper niche for a student as advanced as he, and neither Thomson nor Larmor recognized his genius. Perhaps it was useful to a young man who had been so highly praised at home to find himself ignored abroad. But it was not an easy lesson and intensified his worry about repaying his debts. When, through one of his father's former students teaching in Manchester, Bohr met the university's hearty professor of physics, Ernest Rutherford, he decided to migrate to Manchester to learn something about radioactivity. He arrived in the spring of 1912 and immediately began routine laboratory exercises. Fortunately for physics his radioactive source ran out and so did his patience with the English version of his thesis. He used his new leisure to make contact with a problem with which Rutherford's senior research man, Charles Galton Darwin, was wrestling: how atomic electrons interacted with passing alpha particles. Darwin had simplified the problem by considering the orbiting electrons as free. From calculations Bohr had made for his thesis he knew that Darwin's simplification was illegitimate: the effect on an atomic electron of a passing electrified particle depends on the period of its oscillations around its equilibrium orbit. If the period matches the time of passage, a resonance can occur. That sparked Bohr's interest, since by studying the resonances he might learn something about the period, and hence about the binding, of atomic electrons.

The obvious next step was to investigate the oscillations of the orbiting electrons stimulated by the alpha particle. Bohr soon found himself stymied: the oscillations of the electrons that occur in the orbital plane are unstable; ordinary physics tears apart a nuclear atom containing two or more electrons; the investigation could not proceed as he had planned. That was very agreeable. Failure might point the way: 'it could be that perhaps I've found out a little about the structure of atoms'. Bohr had already recognized that the nuclear atom allowed a clean distinction between ordinary phenomena, which involved its electronic superstructure, and radioactive phenomena, which had their seat in the nucleus. With the information, imparted to him by the physical chemist Georg von Hevesy, who was also in Manchester in 1912, that some substances distinctly different in radioactive properties and in inferred atomic weight were nonetheless chemically inseparable, Bohr worked out for himself the concepts of isotope and atomic number. There was at least as much right as wrong about the nuclear atom.

Bohr might have tried to advance by ransacking the treasure house of spectroscopy, which provided in principle, in its record of the line spectra of the elements, precious information about the vibrations of the aether produced by the oscillations of orbiting electrons. This aether way, from which information about the binding of atomic electrons might be inferred from the aether vibrations they produced, was not the way Bohr chose initially. Instead he adopted the approach to atomic structure pioneered by Thomson, for whom the primary question was how electrons arranged themselves in atoms to produce the periodic properties of the elements. Bohr took up this matter in a memorandum he prepared for discussion with Rutherford in July 1912. Its kernel was a solution to the problem of the size of an atom that had lost all the energy it could by radiation. Just as in the solar system, the principles of ordinary physics do not specify the radii of the orbits; only unknown initial conditions can explain the distances at which the planets circulate around the sun; another sun's planets would take up different positions. But the electronic structure of all atoms of a given element in their ground state seemed to be as alike, as Maxwell used to say, as manufactured articles. Something was missing. The only physical constants implicated in the orbital dynamics of Rutherford's atom were the charge e and mass m of the electron, and the charge on the nucleus Ze, from which no quantity interpretable as a distance could be inferred. Add h, however, and all is well: h^2/me^2 gives a length of the order of atomic dimensions.

Bohr brought h into the nuclear atom by analogy to Planck's resonator. Since, however, the energy of a bound electron is negative in respect of a free one, and the shibboleth $\varepsilon = h\nu$ required a positive quantity, Bohr worked with kinetic energy T. What next? Larmor: 'Attempt to apprehend the exact formal character of the latent connexions between different physical agencies.' Do not fear to leave reality behind! Bohr laid down the implausible condition that in an atom's permanent or ground state the kinetic energy of every electron must be proportional to the frequency of its orbital revolution ω: $T = K\omega$. Owing to the structural differences between a Planck resonator and a Rutherford atom, Bohr left some wiggle room by not specifying K more closely than as a multiple of h. The multiple would turn out to be ½. If the K-condition fixed the size of atoms, it followed that when electrons satisfied it they would be free from the destructive processes expected on ordinary physics (Figure 7).

Bohr thought that his K-condition, in collaboration with the ordinary mechanics with which it conflicted, implied that the extra-nuclear structure of an atom consisted of concentric electron rings; and he was jubilant to discover that energy considerations limited the innermost ring to seven electrons, which, with some good will, might be interpreted as eight, a figure

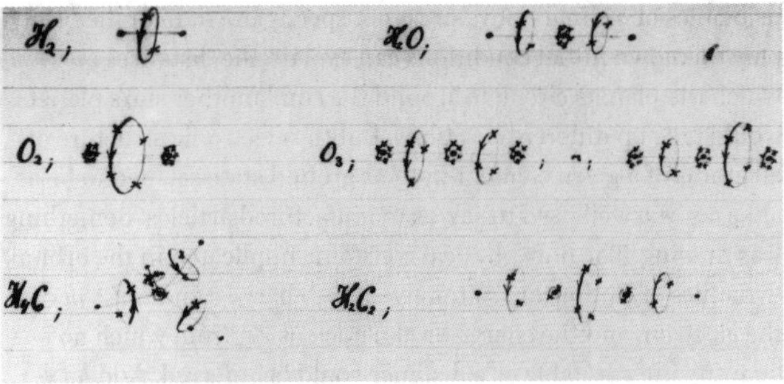

7. **Quasi-realistic diagrams of atoms and molecules from the 'Rutherford memorandum'.**

prominent in the periodic arrangement. '[T]his seems to be a very strong indication of a plausible explanation of the chemical properties of the elements.' The argument is plainly wrong, not because $7 \neq 8$, but because it violates an elementary theorem Bohr proved at the end of the memorandum: the total energy of a ring of symmetrically placed electrons in a circular orbit in a nuclear atom is always equal to the negative of their combined kinetic energy, and so can never be positive. Consequently, ordinary mechanics even when doctored by the K-prescription does not limit the number of electrons in a ring. Perhaps Bohr's hasty error measured his eagerness to bring home decisive evidence that he was on the trail of success.

And in fact he was well started. The coup that put him in command of the world's atomic physicists required navigation around two distinctions between the nuclear atom and a Planck resonator. The frequencies of the periodic motions of bound electrons (M or matter frequencies, designated by ω) must be distinguished from the frequencies of the radiation the electrons produce (A or aether frequencies, designated by ν). Since Planck's resonator directly excites a vibration in the aether at the same frequency as its oscillation, its M frequency ω and the associated A frequency ν are the same. With one essential exception, the equality does not hold for a nuclear atom. The second distinction, closely related to the first, is that whereas a Planck resonator vibrates at the same frequency irrespective of its energy, an orbiting electron must change frequency when it acquires or releases energy. Planck's condition $\varepsilon = h\nu$ applied to the nuclear atom implicates two different M frequencies for every A frequency, one M for the originating, the other for the receiving orbit.

These considerations became pressing for Bohr when early in 1913 a physicist at the University of Copenhagen, Hans Hansen, asked him how his model atom produced spectra. Bohr replied, as he recalled, 'I do not deal with spectra, they are too complicated. The principles of biology cannot be inferred from the colors of butterflies' wings.' Hansen persisted: 'Have you looked at Balmer's formula?' Bohr looked, and saw the light. Balmer's formula gives the frequencies of a series of lines emitted by hydrogen in a way that perfectly represented the relations he sought. The formula, $\nu_n = R(1/4 - 1/n^2)$, where R is a universal constant, related an A frequency to two M frequencies ($R/4$ and R/n^2) and expressed the energy of the emitted quantum, $h\nu_n$, as the difference between two energies ($hR/4$ and hR/n^2). Here the integer n designates the serial number of the line in the hydrogen spectrum; the higher the n the bluer the colour.

Bohr apparently read Balmer's formula in the manner just suggested. Since, as observed earlier, the total energy of an electron in a circular orbit in a nuclear atom is the negative of its kinetic energy T, Balmer's formula interpreted as the difference in energy between the nth and the second orbit gives $T_n = Rh/n^2$. Bohr included this insight in his developing model by generalizing the condition on the ground-state orbit of the Rutherford memorandum, $T = K\omega$, to something like $T_n = K_n\omega_n$. The privileged states of higher energy were capable of radiating. By various precarious arguments Bohr deduced that $K_n = nh/2$ and that an electron could exist in any number of 'stationary' quantized states with kinetic energy $T_n = nh\omega_n/2$.

This generalization came when Bohr read an astonishing series of papers published in late 1912 by a Cambridge mathematician, John William Nicholson. Nicholson traced spectral lines to small oscillations of electrons in a nuclear atom *perpendicular* to their orbital plane, which, contrary to those in the plane, can be stable. Tacitly assuming that the A frequencies emitted in these oscillations are the same as their M frequencies, Nicholson computed the spectra expected from perturbed ring atoms containing four or five electrons. By what now seems sheer coincidence, he found matches accurate to three or four significant figures between the outputs of these rings and no fewer than twenty-four unattributed lines emitted by the sun and certain nebulae. The matches enabled him to fix the radii of the rings, the only free parameters in his model atoms, and so to calculate the energy of the electrons. He expressed the result in terms of angular momentum ($= T/\pi\omega$). In all cases the momenta came out as small integral multiples of Planck's constant h divided by 2π.

Bohr easily checked that electrons satisfying his condition on permissible states had angular momenta equal to integral multiples of $h/2\pi$. He thereupon invented a hybrid model in which an electron approaches a bare nucleus in stages, as an alpha particle becomes a helium atom, dropping from one allowed orbit to another, in each of which it radiated away energy by oscillating perpendicularly to its plane of motion—rather like a pinball falling through a series of tuning forks. The state Bohr considered in his conversation with Rutherford would be the last in the procession of allowed states, reached when the captured electron had lost by radiation all the energy it could.

The hybrid did not last long. When he interpreted the Balmer formula as an energy equation, Bohr jettisoned his compromise with Nicholson and introduced quantum jumps. Taking the running integer n of the Balmer series to indicate the number of an allowable state counting outward from the nucleus, and using his generalized condition on the nth state $(T/\omega)_n = nh/2$,

Bohr had for the nth Balmer line $h\nu_n = T_2 - T_n = (h/2)(2\omega_2 - n\omega_n)$, or $\nu_n = \omega_2 - (n\omega_n)/2$.

In words, the A frequency of the nth Balmer line is the M frequency of the second allowed orbit less one-half n times the M frequency of the nth orbit! The A frequency relates only in an indirect and opaque manner to the M frequencies of the electron responsible for its production. This is the feature of Bohr's quantum atom that seemed the most extraordinary to the best informed and most critical of his contemporaries because it replaced the ordinary theory of light emission by continuous vibrations with an unintelligible postulate of unanalysable jumps.

Rutherford's reaction is instructive. Unable to picture how an electron could stir up the aether without vibrating, he objected that it would have to know where it was going to stop before it jumped so as to vibrate appropriately during the journey. And how could it know its terminus in advance? Bohr said that quantum laws do not allow us to pry into an electron's travel plans. Einstein's reaction was amazement according to Hevesy, who reported it with creative orthography: 'he was extremely astonished and told me "than the frequency of light does not depand at all on the frequency of the electron...this is an enormous achiewement."'

And what could Bohr offer as persuasive confirmation of his bizarre ideas? He could compute the constant R in terms of more fundamental quantities: e and m, which characterized the radiator, and h, which, with e and m, defined the size of the atom. The theoretical expression agreed with the empirical value of the hydrogen constant R_H to around 6 per cent. Spectroscopists, who customarily worked to an accuracy of five or more significant figures, were not impressed and pointed to another series of lines they attributed to hydrogen for which the agreement was even poorer. This series followed a Balmer-like expression with half-integer terms: $\nu_n = R[1/(3/2)^2 - 1/(n/2)^2]$. Half integers being impossible in his theory, Bohr rewrote the pseudo-Balmer formula as $\nu_n = 4R[1/3^2 - 1/n^2]$ and attributed the lines it represented to ionized helium, for which $Z = 2$ (R contains the factor Z^2).

Spectroscopists looked in tubes of helium purged of hydrogen and found the lines, thus correcting a serious misattribution. Bohr's coup instantly raised a significant new objection, however: the empirical value of the constant for ionized helium (R_{He}) was not four times hydrogen's as his theory required, but 4.00163. Bohr coped with this emergency with a masterly retreat to classical mechanics. He had supposed the nuclear mass to be infinite in comparison with the electron's; but in fact it is only 2000 times greater for H and 8000 times greater for He. Introducing these numbers by the ordinary method of orbital mechanics, Bohr made $R_{He}/R_H = 4.0016$. (Since factors in e and h cancel out in the ratio, the considerable contemporary uncertainties in their values did not matter.) The improvement in Bohr's theory precipitated by the fruitful objections of spectroscopists pointed the way forward. Henceforth, he would develop mechanical models with the utmost faithfulness before compromising them with quantum concepts.

The quantum side of the theory, however, the basic postulate on which numerical agreement depended, $T_n = n\omega_n h/2$, had only the justification of success. Bohr cast about for a foundation. The harvest was four different and mutually inconsistent, or barely consistent, groundings. The first two developed different analogies to Planck's theory, supposing the emission of T_n to occur during the capture of a free electron into the nth stationary state. Then, with an eye to Planck's postulate and the defining equation of the stationary state, $h\nu_n = T_n = n\omega_n h/2$, Bohr had the enigmatic result, $\nu_n = n\omega_n/2$. Did this mean that the A frequency is $n\omega/2$ or, if the radiation occurs in n steps, $\omega/2$? Bohr did not know and offered both possibilities. And why did half the M frequency appear? Bohr suggested that the A frequency might be an average of the M frequencies of the participating orbits; taking 0 for that of the unbound state, the average is $\omega/2$. The suggestion seems ad hoc, as it may have been; but it had the merit of consistency with the queer result that two M frequencies are needed to give rise to one A frequency.

The third of the justifications of the defining equation $T_n = n\omega_n h/2$ is an adumbration of the Correspondence Principle ('CP'), the method Bohr would develop to explore his quantum labyrinth. In this preliminary phase, the CP required that in a transition between neighbouring orbits n+1 and n, where n is much larger than 1, the A frequency must approach numerical equality with the M frequency. But there are two M frequencies! In the limit of correspondence, however, they become asymptotically equal. At large values of n, $\nu_{n+1,n} = R[1/n^2 - 1/(n+1)^2] \approx 2R/n^3 = \omega_{n+1} \approx \omega_n$. Bohr then had $R = n^3\omega_n/2$. An algebraist not bothered by contradictions could easily obtain an expression for ω_n. One needed only to eliminate a_n, the radius of the nth orbit, between ordinary physics' equation for circular motion, $Ze^2/a_n^{\ 2} = m\omega_n^{\ 2}a_n$, and Bohr's fundamental quantum hypothesis, $T_n = m\omega_n^{\ 2}a_n^{\ 2}/2 = nh\omega_n/2$. The manipulation recovered Bohr's earlier value for R, which deserves to be recorded in all its beauty here: $R = 2\pi^2 mZ^2e^4/h^3$.

These three derivations—the two from Planck's concepts and the one from the proto CP—may be considered aether formulations as they derive from conditions on the radiation. The fourth derivation returns to the matter approach of the Rutherford memorandum and places a condition on the orbits: the angular momentum p_n of an electron in its nth allowed circular path equals $nh/2\pi$. Bohr soon dropped the two Planck derivations as misleading, took the third as the basis of his subsequent principled derivations, and used the fourth for convenience when discussing circular orbits despite its mixing of ordinary mechanics and quantum ideas. Though altogether misleading, and it misled many, the condition on the angular momentum proved fruitful. As Bohr put it in his oracular way, 'While there obviously can be no question of a mechanical foundation of the calculations given in this paper, it is, however, possible to give a very simple representation of the calculation...by [the] help of symbols taken from the ordinary mechanics.'

These several attempts to ground his novel atomic theory give a precious insight into Bohr's mind at work, into his way of entertaining several contradictory formulations of his thought at the same time. The method required what Einstein called Bohr's 'unfailing tact', his capacity to choose just those footholds from which he could safely advance, moving slowly from one shaky position to a marginally better one, and then, perhaps, forsaking them all. Long after his orbital model had been superseded by quantum mechanics, he was asked about the arguments he had used to formulate it. He replied that he could not have intended them seriously, although later in the same interview he recalled defending every word of the final draft of the paper as 'quite essential to the argument' when Rutherford offered to cut it down. What then was the justification of the averaging to get the factor ½ in the fundamental equation $T_n = nh\omega_n/2$? 'That was just the stupidity of the way of looking at it.' What about the odd analogies to Planck's theory? 'That is taken too seriously, you see. It's not so, actually... It was not taken seriously at all. There are some sentences about this which I actually agree are nonsense.... It is hard for me to see what it means.' What about the condition on the angular momentum? 'It really would have been much more beautiful if it had all been left out.' And the entire approach? 'Most of it is sheer nonsense.'

Unfinished business

'Shall we really try?' While yet in Manchester defending every word of Part 1 of the 'Trilogy' into which he had expanded the Rutherford memorandum, Bohr asked Margrethe whether she was up to Parts 2 and 3. These dealt with the tremendous question Bohr had considered before meeting with the Balmer formula. Now he had an exact *K*-principle, which he took from Part 1 in its fourth formulation: in the ground state of an atom, *every* electron has angular momentum equal to $h/2\pi$. To exploit this formulation to explain the periodic properties of the elements, Bohr returned to the tricky question of the mechanical stability of the unstable rings. He ignored the suicidal oscillations in the plane of the rings and argued from the conditionally stable oscillations perpendicular to the rings that Nicholson had studied. The chief result of his classical computations was that a single ring of n electrons orbiting a charge ne can be stable against perpendicular displacements if $n \leq 7$ and every electron's angular momentum around the nucleus remains $h/2\pi$. This recovered the result claimed in the Rutherford memorandum through an error in calculating potential energy!

With this result now secure, Bohr proceeded to assign ring structures to elements up to $Z = 24$ (chromium) without bothering with the number 7. As he admitted, his principles usually did not determine the electronic distributions unambiguously. When stuck, he appealed to a mixture of delicate mechanical arguments and the brute facts, as represented mainly by chemical valence.

In his usual manner, he considered the formation of an atom by successive capture of Z electrons by a bare nucleus of charge Ze. As we know, ionized helium behaves like hydrogen; but when it adds its second electron, forming a ring, neutral helium holds together tightly, being a little over half the size of a hydrogen atom. Lithium presents a problem. It binds its first two electrons more tightly than helium's and would do the same with the third if nature followed Bohr's requirement that the ground state should possess less energy than any other configuration subject to his restriction on angular momentum. But the chemical evidence implied that lithium has one easily detachable electron. Hence Bohr adopted the structure he denoted, in obvious shorthand, 3(2,1). Similar considerations produced for beryllium the arrangement 4(2,2).

These prescriptions were unprecedented. Although Thomson had indicated connections between chemical properties and certain arrangements of the rings in his atom, he never specified the structure of any. And although Bohr could not go beyond helium on a principled basis, his revived confidence, fed by Margrethe, Rutherford, and praise for Part 1, carried him where no one had been before. When he reached the watershed of beryllium, he stopped to examine the behaviour of two concentric rings containing the same number of electrons. Invoking a force to pull the inner ring upward parallel to itself, and so to displace the outer one downward, Bohr supposed that the inner would grow and the outer contract until they come to the same size with the electrons in one opposite the gaps between electrons in the other. If the extraneous force now disappears, the rings may coalesce. It appears that 'there is a greater tendency for the confluence of two rings when each contains the same number of electrons'. Hence, as electrons are added to build up atoms beyond beryllium, its two inner rings of two electrons each would tend to coalesce into one of four.

As the intrepid explorer proceeded further into unknown territory, he expected to find landmarks where two four-rings flow into one of eight, for which, as suggested by the length of the second and third periods of the table of elements, nature has a particular partiality. Just when two two-rings unite into one four-ring, or two four-rings into an eight, 'cannot be determined from the theory'. With an eye to chemistry, which, as in the case of lithium, gives a strong clue to the number of electrons in the outermost ring, Bohr increased the innermost (the first) ring to four at nitrogen, 7(4,3), and from four to eight at neon, 10(8,2). Meanwhile, the two outer two-rings of oxygen, 8(4,2,2), flow together in fluorine, 9(4,4,1). The innermost ring remains at eight from neon to chromium, to which Bohr assigned the structure 24(8,8,4,2,2). 'Without any further discussion it seems not unlikely that this constitution of the elements will correspond to properties of the elements similar with those observed.' In particular, the building out of internal rings explains the similarity of chemical properties among the transition metals, and again among the rare earths. Bohr was quite pleased with Part 2, in which, he touted, 'the application of Planck's theory of radiation to Rutherford's atom-model through the introduction of the hypothesis of the universal constancy of the angular momentum of the bound electrons leads to results which seem to be in agreement with experiments'. And yet all his guesses beyond carbon, 6(2,4), were wrong.

X-rays exposed his errors. His contemporary at Manchester, Henry Moseley, discovered that a formula much like Balmer's applied to the highest frequency lines atoms emitted. The most intense of these lines, arbitrarily designated K_{α}, satisfied the equation $\nu_K = R(Z-1)^2(1/1^2 - 1/2^2)$ for the elements from calcium ($Z = 20$) to zinc ($Z = 30$). Bohr followed Moseley's work closely and probably had a hand in its design, for Bohr was present in Manchester in July 1913 to discuss Parts 2 and 3 with Rutherford just at the time that Moseley was planning his survey of the X-ray line spectra (Figure 8).

Two features of the simple equation in which Moseley encapsulated his results surprised both of them. For one, there was no indication of periodicity and thus, since Bohr supposed the K lines to originate from the innermost ring, no evidence for the coalescence he had expected. Secondly, they could not account for the Balmer-like formula. The principles of Part 2 blocked the obvious explanation, subsequently adopted: the innermost ring has no more than two electrons in the ground state; removing one leaves an effective nuclear charge of $Z-1$; the electron that fills the vacancy from the second ring has two quanta of angular momentum. Bohr's analysis of the coalescence of rings gave the innermost eight electrons beginning below calcium and his basic principles restricted all electrons in the normal atom to one quantum of momentum. He had painted himself into a corner. 'For the present [he wrote Moseley] I have stopped speculating about atoms.'

8. **Bohr and Rutherford returning from a walk and talk in the mud.**

Part 3, the shortest part of the Trilogy, centres on the stability of two nuclei, usually taken as identical, held together by a belt of electrons. Calculations similar to those in Part 2 showed that the dumbbell system can only be mechanically stable under quite

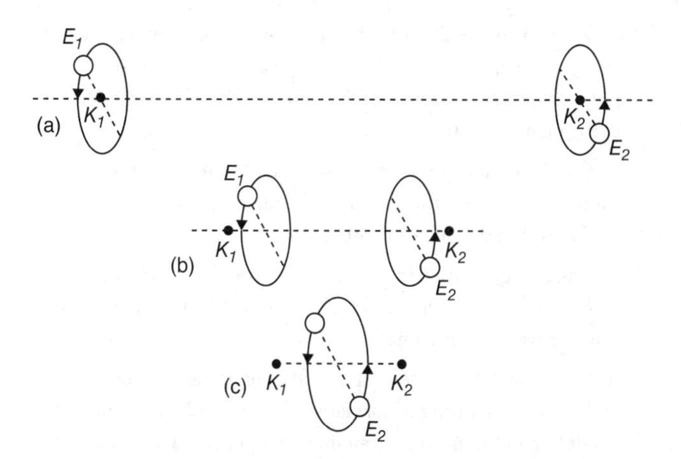

9. The formation of a hydrogen molecule.

narrow criteria, which Bohr applied to covalent bonds between atoms with a valence of one or two. Proceeding as usual, he imagined assembling the molecule from its atomic constituents. In the simplest case, extraneous forces constrain two hydrogen atoms to approach one another with their rings parallel and the orbital motions of their electrons 180 degrees out of phase. Bohr said that he had done calculations that proved the stability of the motions throughout the process and that no extraneous forces needed to act since the two atoms attract one another (Figure 9).

Beyond hydrogen, Bohr offered arguments against the existence of a helium molecule and played with models of HCl and H_2O held together by electron rings. To go further would have been, as Bohr remarked, to fly 'far out of the range' of his theory. In fact, he never tried the flight again. He concluded the Trilogy, his great contribution to modern physics, with a restatement of its principles, five in all. Some he did not follow scrupulously and others he soon abandoned.

1. Radiation is not emitted or absorbed continuously but only when the system passes from one stationary state to another—a proposition Bohr violated for disturbed oscillations perpendicular to the plane of motion.
2. Ordinary mechanics govern the stationary states but not the transitions between them—also violated, for the outlawed oscillations in the plane of motion.
3. The frequency emitted in a transition between stationary states involves a change of energy $h\nu$. The emitted radiation pursues its way as a wave in the aether.
4. A stationary state is determined by the condition that the total energy emitted during its formation divided by the frequency of revolution of the electron is an integral multiple of $h/2$, which, for a circular orbit, makes its angular momentum around the nucleus an integral multiple of $h/2\pi$—a formulation that fuses or confuses a condition on radiation with one on mechanical motions.
5. In the 'permanent' or ground state of a ring atom, every electron has exactly one quantum of angular momentum—a restriction soon abandoned.

Bohr had planned to treat magnetism in what would have been Part 4 of his waxing treatise on the constitution of atoms and molecules. The subject seemed ripe: his stipulation that each bound electron must conserve its angular momentum even when moved into a magnetic field suggested a way to circumvent the criticism that he had levelled against the classical theory of dia- and para-magnetism in his thesis. Two big snags lay in the way. For one, classical electrodynamics associates with the angular momentum of an orbiting electron a magnetic moment, which, when quantized by Bohr's rules, came out to be five times larger than measurement allowed. The other snag was that Bohr's radiation by quantum jumps cancelled a theory that had been awarded a Nobel prize. The theory explained what became known as the 'normal Zeeman effect' after Pieter Zeeman's detection of the splitting of a spectral line into three components by a magnetic field. Its explanation, worked out by Zeeman's professor Hendrik Lorentz, could not be squared with the nuclear atom. Lorentz had treated the electrons responsible for the line as Planck resonators and, naturally, had assumed the identity of M and A frequencies. Bohr had no quantum alternative in hand.

Chapter 3
Magic wand

While Bohr struggled with ambiguity, several lines of work independent of his confirmed the fertility of the nuclear model and his approach to it. At the same time, experiments inspired or guided by his ideas gave reassurance that, if he could keep marching with two feet in the air, he would lead the attack on the microworld. Although small in number, the international brotherhood of physicists working on projects relevant to Bohr's programme produced enough pertinent information during the year before mobilization for the Great War to keep him occupied. He gave what time he could to nourishing his atom while fulfilling his new family obligations to Margrethe and his teaching obligations, primarily lecturing to medical students, at the Polytechnic Institute.

The coming of war gave Bohr breathing space by drawing many colleagues and competitors in the belligerent countries into military service. Darwin was one of them. When he went to war, in 1915, Rutherford called Bohr to replace him. During this second Manchester interlude, which lasted a year and a half, Bohr's friends in Copenhagen lobbied the university to establish a professorship in theoretical physics for him. Among the most effective of the lobbyists were Hanna Adler and Valdemar Henriques, a student of Christian Bohr and his successor as physiology professor, 'the most truly faithful friend of the whole [Bohr] family'. Henriques raised support from Jewish philanthropists to acquire land for the Institute for Theoretical Physics that Bohr would head and, as chairman of the board of the Carlsberg Foundation in the 1920s, provided subsidies to equip it.

Bohr returned to Copenhagen in 1916 to continue his efforts to clarify the fruitful muddle he had made and to improve his working conditions. As his new professorship came with a room scarcely large enough to house his ideas and no laboratory, he set himself to improve the physical world in two different ways simultaneously. In 1921 his Institute opened for business on the edge of a park north of the city centre, and in 1925 efforts he spearheaded arrived at a consistent formulation of the quantum physics of the atom.

Instant progress

Chief among the corroborating investigations inspired by the Trilogy completed before the war was the untangling of the hydrogen–helium spectrum. The experimental work, by Evan Evans, took place in Rutherford's laboratory in 1913. Another corroborating investigation, Moseley's work on X-ray spectra, produced the tantalizing results already mentioned. A third investigation, which owed nothing in its origin to Bohr's theory, gave powerful support to it—but only after Bohr showed that the investigators had misinterpreted their results. James Franck and Gustav Hertz, then junior researchers at the University of Berlin, measured what they took to be the minimum energy necessary to ionize (remove an electron from) a gas molecule. Their measure of ionization was the velocity at which a beam of electrons passing through a gas began to transfer energy to its molecules. That the criterion was apt seemed demonstrated when they detected a positive current, presumably of ionized molecules, flowing toward the negative electrode in their experimental apparatus. Since the consequent value for the ionization energy fell far below what Bohr could estimate from his quantum atom, they decided that they had refuted it.

That set up a victory as striking as the triumph over the hydrogen–helium spectrum. Franck and Hertz had not produced ionization, Bohr answered, but only excitation; they had lifted an outer electron to the lowest unoccupied stationary state, not knocked it out of the atom. Whence then the positive current they had found? From the radiation emitted in the return of the excited atom to its ground state! This radiation, being in the ultraviolet, created a photo-effect, that is, ejected an electron, on striking the cathode: what Franck and Hertz identified as a positive current towards it was in fact a negative one from it. This reinterpretation, which established a threshold for a free electron to yield energy to a bound one, provided evidence other than spectroscopic for the existence of stationary states. It brought Franck and Hertz the Nobel Prize in 1925. By then their experiment was so tightly integrated into Bohr's quantum atom that Franck forgot that he had begun by doubting it.

A final case, which likewise came as an unanticipated gift, was the obscure phenomenon detected by Johannes Stark, who marched loudly to his own drummer. According to ordinary physics, spectral lines were immune to splitting by an electric field of the strength Stark commanded. Nonetheless he showed that it could split Balmer lines. Reasoning that Bohr's atom might allow what the old physics prohibited, several bold physicists added a little electrostatic energy to his equations and managed to get rough agreement with Stark's numbers. None bothered to work out how the applied field would distort the circular Balmer orbits. Perhaps it did not seem to them worth the effort to investigate a clear classical problem in detail as a preliminary to infecting its solution with quantum concepts. Bohr in contrast solved the classical problem as best he could so as to be able to apply the embryonic CP in a disciplined way. He then calculated to reasonable accuracy the frequencies of the strongest two of the five satellites into which Spark split the Balmer lines.

Still the repertoire of models accessible to a correspondence treatment remained small. How to enlarge it? For a time, Bohr favoured a way to transform a case he could handle into a new one without changing its quantum state. An enticing example would be transforming a nuclear atom into a Thomson model by supposing that the nucleus gradually swells to engulf the orbiting particles. (In Thomson's model the electrons circulate in rings *within* a space that acts as if it were a nebula of positive electricity.) A professor at the University of Leiden, Paul Ehrenfest, who would become a great fan and friend of Bohr, had generalized a classical theorem to secure the continuity of the quantum state of hydrogen's electron during such a process. This 'adiabatic' theorem declares that the quantity $I = 2T_{ave}/\omega$ remains constant during very slow changes in the environment; for example, I remains constant for a simple pendulum as its length is shortened over a time long in comparison with its period.

The Is associated with periodic motions thus made excellent candidates for quantization. For Planck's resonator, where $2T_{ave}$ equals the total energy, the theorem suggests $I_n = \varepsilon_n/\omega = nh$. For the nuclear hydrogen atom, attainable adiabatically from the resonator, $(T_{ave})_n = nh\omega/2$, just Bohr's quantum condition on the stationary states. This approach lay behind Bohr's analysis of the Stark effect, in which a Balmer circle is slowly pulled into a very eccentric ellipse, changing its energy without altering its quantum state.

Bohr had a paper in press on adiabatic transformations and other fundamental issues when in the early spring of 1916 he received in England via neutral Denmark some articles by Arnold Sommerfeld, professor of physics at the University of Munich.

Though belligerent enough for war service, Sommerfeld was old enough to avoid it, and had devoted his ongoing research to mathematical extensions of Bohr's ideas. A glance at Sommerfeld's articles was enough to send Bohr post haste to London to pull his paper from the publisher. He had at least two reasons to draw back. For one, Sommerfeld had found a formal way to quantize systems with more than one periodic motion and had demonstrated the fertility of his method in several striking ways. For another, Bohr did not like the formality of Sommerfeld's approach, which avoided a close analysis of the classical motions before quantization, but he did not know how to replace it.

Sommerfeld's formal enlargement of Bohr's scheme started with the condition on the stationary states, which he wrote as $2\pi p = nh = 2T_{ave}/\omega$ in order to be able to handle periodic motion with non-constant kinetic energy, as in an ellipse. The enlargement consisted of making this condition into two, one for rotation (the motion in azimuth), the other for pulsation (the change in radius from apogee to perigee). Since the energy of an electron moving in an ellipse depends only on the length of its major axis and not on its eccentricity, Sommerfeld's introduction of a second condition just gave back the result Bohr had obtained for a circle, $T_n = Rh/n^2$.

The power of Sommerfeld's method appears where the attractive force on the electron cannot be assimilated to that operating between planets or point charges. This situation arises in the hydrogen atom in its ground state because its electron travels fast enough (at 1/137 the speed of light) to show relativistic effects. The increase of mass with velocity claimed by the theory of relativity prevents the closure of elliptical orbits; the major axis revolves at constant angular velocity around the nucleus with a frequency that depends upon the eccentricity (Figure 10). Sommerfeld's conditions fix the possible values of the eccentricity of an elliptical orbit whose major axis is determined by the quantum number n as $\sqrt{(1 - k^2/n^2)}$, where the number k quantizes

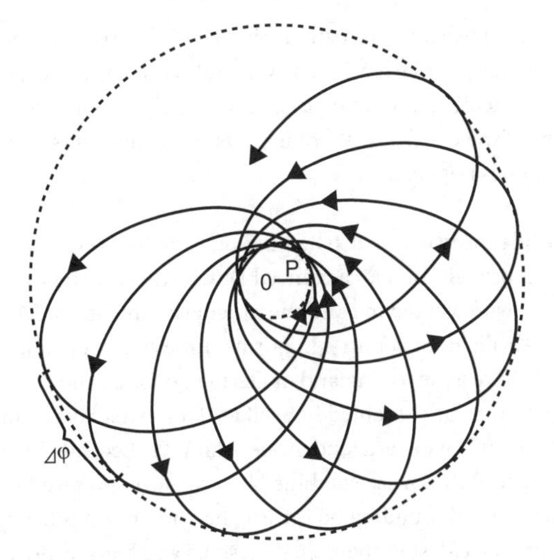

10. A precessing elliptical orbit.

the azimuthal motion. Relativity makes n ellipses with different energies on the same major axis ($k = 1,2\ldots n$). When tested experimentally with ionized helium, where the relativistic effect is much larger than in hydrogen, Sommerfeld's theory proved only too good. His calculations agreed well with the lines observed; but no lines appeared for transitions in which k changed by more than one unit.

The problem of the k restriction became more pressing with the conquest of the Stark effect by Paul Epstein, a Russian Jew confined to Sommerfeld's institute as an enemy alien. Epstein found that the Stark transitions, which he required three quantum numbers to describe, were no freer than he. What constrained the jumps? Another Jewish wartime detainee in Sommerfeld's institute, Aldebert Rubinowicz, had an answer. According to Rubinowicz's calculations, ordinary theory did not allow an electromagnetic wave of energy $h\nu$ a greater angular momentum than $h/2\pi$; thus k could not change by more than 1, but as the argument allowed it to remain unchanged ($\Delta k = 0$), it did not agree perfectly with experiment. When Bohr digested the output of Sommerfeld's quantum concentration camp, he enlarged the purview of the CP. One by-product was ruling out transitions in which k does not change.

Meanwhile the war was also helping Bohr to develop a research school. It brought him an advanced Dutch student, Hendrik Kramers, who wanted to continue his education outside Holland in a non-belligerent country. In normal times he would almost certainly have gone to England or Germany; under the circumstances, he presented himself at Bohr's small office and asked to be taken on as a student-assistant. He became Bohr's Epstein and Rubinowicz combined and replaced Margrethe as amanuensis and sounding board. She had other things to do: Christian, the eldest of the Bohrs' six sons, was born in 1917. At first Kramers lived literally on beer money, on a grant Bohr obtained from the Carlsberg Foundation. He soon produced an extraordinary thesis.

'Hanging in language'

By 'hanging in language' Bohr meant 'dependent on unambiguous communication'. His atom was anything but unambiguous. At some point he realized that to describe the microworld physicists needed a new syntax, not yet glimpsed, which would combine quantum quantities as ordinary physics did their analogues. Bohr had identified one pair of analogues, the frequencies $\nu_{n,n\text{-}1}$ and ω_n. Kramers took an essential step forward by working out a correspondence of intensities, which, with an intervention by Einstein, enabled identification of a second set of analogues.

In keeping with Bohr's methods, Kramers began with a thorough analysis of the classical description of the emission of radiation from a charged particle moving periodically. His measure was

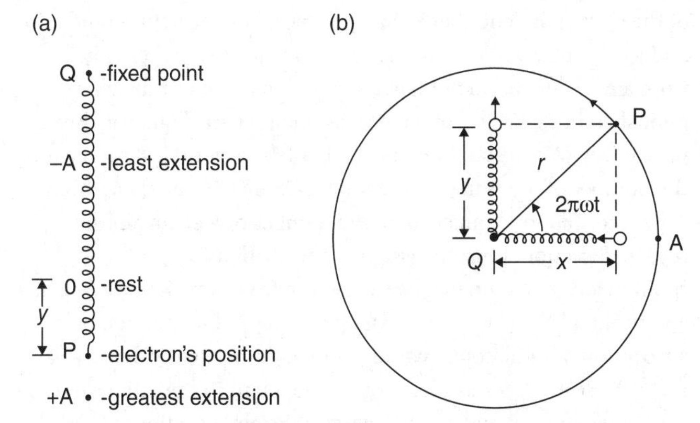

11. Orbits and resonators: (a) the extension y of the 'spring' of a Planck resonator is $y = A\sin 2\pi\omega t$; the frequency ω is proportional to the square root of the constant defining the spring's strength; (b) a circular orbit decomposed into resonator vibrations; displacement along x is $r\cos 2\pi\omega t$ and along y is $\sin 2\pi\omega t$ if the motion begins at A.

intensity, the energy emitted per second, and his model a collection of Planck resonators. The intensity of radiation from a resonator is proportional to the square of its amplitude A. Circular motion presents a similar situation, since the rotation can be compounded from two resonators oscillating at right angles at the same frequency ω but 90 degrees out of phase (Figure 11). The ellipse brings something new. Owing to greater irregularity in the rotation, a pair of resonators does not suffice; indeed, an infinite set, most with vanishingly small amplitudes, is required. For a closed ellipse, each of them must vibrate at an integral multiple, or overtone, of the fundamental frequency involved, at ω, 2ω, 3ω...in order that the electron return to the same place in the period 1/ω.

The situation required new words and symbols. $A_\tau(n)$ is the generic one: it designates the relative strength of the τth overtone in the set of oscillators representing elliptical motion with fundamental frequency ω_n. Using this language, the generalization of the correspondence between ω_n and $\nu_{n+1,n}$ to one between $\tau\omega_n$ and $\nu_{n+\tau,n}$ is irresistible. The wanting analogue to $A_\tau(n)$ emerged from a new derivation of Planck's radiation law that Einstein published in 1916. The analogue was not a physical quantity but a probability. When stability sets in (should the world last so long), the number of molecules in the walls of Planck's oven going from state n to state $n + \tau$ must equal the number proceeding in the reverse direction. Einstein assigned a probability $a_{n+\tau,n}$ of spontaneous emission by quantum jump from state $n+\tau$ to state n, and probabilities $\rho(\nu_{n+\tau,n})b_{n+\tau,n}$ and $\rho(\nu_{n,n+\tau})b_{n,n+\tau}$ for emission or absorption of radiation in equilibrium with the energy density $\rho(\nu_{n+\tau,n})$. He accepted as measures of the relative populations of the stationary states n and $n+\tau$ the classical formula for the distribution of energy among gas molecules in equilibrium, and inconsistently set the energy difference of the states equal to the quantum difference $h\nu_{n+\tau,n}$. The expression for ρ obtained by equating emitted and absorbed radiation was, astonishingly, Planck's radiation formula. The derivation required a new physical concept, 'stimulated emission', expressed by the term $\rho(\nu_{n+\tau,n})b_{n+\tau,n}$. It is the theoretical principle of the laser. The recovery of Planck's formula persuaded Bohr that Einstein's unusual treatment contained some measure of truth. It was therefore a 'natural generalization' (as Bohr would say) to take Einstein's probability coefficients $a_{n+\tau,n}$ to be the atomic analogues of the classical $A_\tau(n)$.

That left the unknown syntax connecting the νs and as as a research project. If found, it would bypass hypothetical electron orbits in favour of observable quantities, the frequencies and intensities of spectral lines. Bohr's remark at the end of Part 1 of the Trilogy that mechanical quantities like momentum defining the orbits had only symbolic value may apply here. In classical physics the equations that describe heat and current flow have the same form, temperature in the one case being the analogue of electric potential in the other. The equations might therefore be taken as primary symbols, and the theories of heat and electricity two of their instantiations. Bohr construed his model with its undetectable electron orbits as an imperfect instantiation of the symbols properly descriptive of the quantum world.

Kramers extended his analysis to the open ellipse that Sommerfeld had invented by imposing relativity on ionized helium. Now the representation of the motion had to include resonators with the frequency of the precession, ω_k. Classical analysis revealed that only resonators of frequencies $\tau\omega_n$ ffl ω_k were required; the amplitude for a transition in which k did not change was zero. Consequently, in the correspondence limit, transitions with $\Delta k = 0$ do not occur. Bohr supposed that the prohibition held throughout the atom and so could eliminate the possibility of the unobserved transitions that Rubinowicz had to admit. To Sommerfeld this was abracadabra and the CP a magic wand; but since it appeared to work only in Copenhagen, he went on adding quantum numbers to classify spectra unconcerned about the finer mechanical motions in play.

The way of the CP was arduous even where it worked. In 1918, when engaged with the second part of a lengthy and difficult work intended to reformulate the Trilogy so as to give Sommerfeld's methods a deeper foundation, Bohr wrote to an English colleague, Owen Richardson, about his manic-depressive scientific life. He experienced 'periods of overhappiness and despair, of feeling vigorous and overworked, of starting papers and not getting them published, because all the time I am gradually changing my views about the terrible riddle which the quantum theory is'. There was a physical as well as a psychological price to pay: constant fatigue that prompted cessation of work on doctor's orders in 1921.

The work under way when Bohr wrote to Richardson privileged adiabatic invariants as quantities suitable for quantization, and the CP, expanded to include Einstein's coefficients, as the guide to quantum equivalents of classical quantities and syntax. But although Bohr's new presentation exploited elegant formulations of orbital mechanics used by astronomers, it did not bring quantitative success beyond hydrogen-like atoms. Neither he nor Kramers could calculate the emission spectrum of normal helium in agreement with experiment. As they were failing to conquer helium and the few physicists who read his revised principles, which he published in an inaccessible Danish journal, were failing to understand them, Sommerfeld brought out the first edition of his *Atombau und Spektrallinien* (1919), which applied his formal techniques successfully to a wide range of problems.

Sommerfeld described his text as a rendition of the 'music of the atomic spheres' (Figure 12).

Bohr did not like the score. It had too much of formal mathematics and too little of foundational probing for his taste. But he could not compete with Sommerfeld in extending analysis

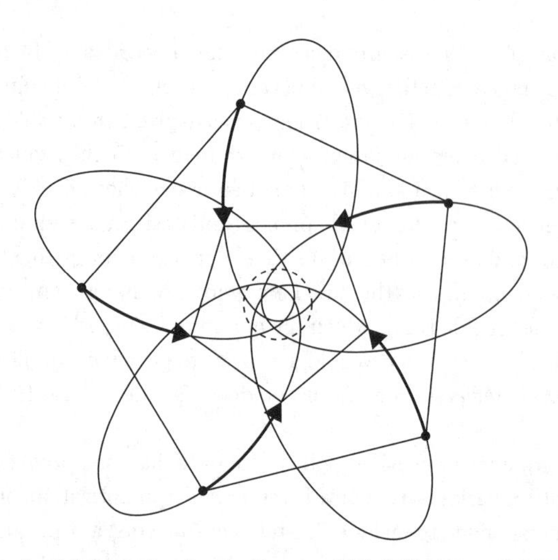

12. Music of the atomic spheres: Sommerfeld's 'Ellipsenverein', in which a chorus of electrons simultaneously traverse tuned elliptical paths.

to complicated spectra or in providing the clear pedagogical guide to the field that successive, expanding editions of *Atombau* offered. Instead he returned to 'Atombau' full stop, to the problem that had drawn him to the quantum atom in the first place: accounting for the periodic properties of the elements. He now had two quantum numbers to work with, the azimuthal k defining angular momentum as well as the original n defining energy. His new analysis, completed, provisionally, in 1922, brought him two spectacular acquisitions. One was a new element, which chemists had missed although it had been in plain sight for decades on Mendeleev's table. The other was Sommerfeld's prize student Heisenberg, who would come to the Institute for a finishing touch. His collaboration with Bohr and Kramers set up his invention of matrix mechanics.

The Bohrfest and the Nobel Prize

To add to his and ups and downs, Bohr took on the task of planning the fabric of his Institute just when he was struggling with his new principles of atom building. The government agreed to ground breaking on 1 November 1918. Ten days before the Armistice, Bohr's building, which he would use to promote internationalism as well as science, jumped from speculation to materialization. Although construction had its difficulties owing to inflation, labour trouble, and Bohr's many changes in plan, the building was ready for occupancy in January 1921 (Figure 13). He then occupied it, quite literally, as a third of it compromised living quarters for his family; there were also an apartment for a mechanic, a suite for visitors, studies for himself and Kramers, a library, and rooms and workshops for experiments. The fabric cost twice Bohr's estimate of 1917, the experimental equipment thrice.

The Carlsberg Foundation furnished the most important piece of apparatus, a spectroscope ordered from England; in justifying its expense, Bohr submitted a letter pointing out that because the war had devastated German universities and the harsh peace made it

13. The Bohr Institute in 1921.

impossible to rebuild their experimental apparatus, Denmark should seize the opportunity to become 'an international place of work for foreign talent whose own countries are no longer in a position to make available the golden freedom for scientific work'. The author of this letter was not Bohr but Sommerfeld, who visited magical Copenhagen in September 1919 when Bohr was struggling to find funds for equipment. The Carlsberg Foundation understood the argument, as did the Danish government, which in October 1919 set up the Rask–Ørsted Foundation to support foreigners doing research in Denmark. The coupling of the names of a philologist, Rasmus Rask, and a physicist, Hans Christian Ørsted, in a research foundation was as novel as its purpose.

Bohr tapped Rask–Ørsted money to bring the first foreign research workers to the Institute apart from Kramers: a Swede, a Norwegian, and a Hungarian (Bohr's old friend Hevesy). In all, the Rask–Ørsted Foundation supported thirteen foreign physicists to work at the Institute and a number of short-term visitors, who may have included the first of the foreign experimentalists, James

Franck. To complete his staff and assure its proper functioning, Bohr hired a competent secretary, Betty Schultz, who remained with him for the rest of his working career.

In the summer of 1922 German physicists showed their appreciation of Bohr's scientific and political achievements by holding a meeting in his honour at the University of Göttingen, whose new professor of physics, Max Born, had taken an interest in the quantum theory of the atom from an angle more mathematical than Bohr's. The several lectures Bohr gave there, full of exciting novelties and oracular hints, surprised an audience unused to approaching physics philosophically. Bohr ended his Göttingen lectures by pulling another rabbit from his hat. The trick began with assigning to each electron of every known atom its particular values of the quantum numbers n and k.

Bohr claimed that the CP waved away such structures as Sommerfeld's elegant *Ellipsenverein* and required instead that in their ground states the electron orbits be interlaced in three dimensions so as to achieve the lowest total energy compatible with the quantum restriction on k. As before, he imagined each atom to be built up through the successive capture of free electrons by a bare nucleus. Evidently, hydrogen's single satellite must be in a 1_1 circle, since for $n = 1$ the only possibility is $k = 1$. Helium requires two equivalent orbits: its electrons share a circle. Chemical and spectroscopic evidence place the third electron's, lithium's, in a loosely bound 2_1 ellipse. For want of an alternative, Bohr assigned the next three electrons to 2_1 orbits also and the next four to 2_2 circles. That made neon's structure $(1_1)^2(2_1)^4(2_2)^4$, where the exponents signify the numbers of electrons in the n_k state (Figure 14).

The third period repeats the second, thus (argon) = (neon) $(3_1)^4(3_2)^4$. The fourth opens with two 4_1 orbits (for potassium and calcium), which penetrate deeply within the atom; all this

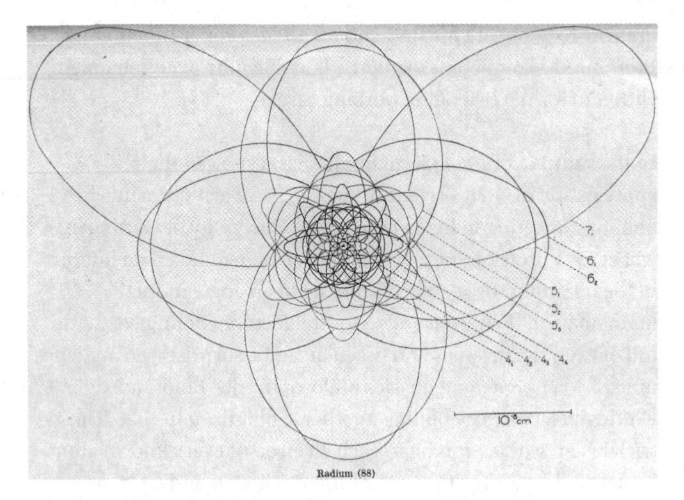

14. The eighty-eight orbits of radium, each labelled by its n_k designation in Bohr's *Atombau* of 1922. In the original drawing, which Bohr commissioned though he did not believe in literal orbits, paths with even values of n appear in black, with odd values in red.

supposedly secured by heroic calculations of the mechanical response of previously bound electrons to a new arrival and the application of the CP. With the third element of the fourth period, scandium, something new occurs, a series of 'transition metals' that differ chemically from one another much less than successive elements of the second or third period do. Bohr accounted for them by adding ten more 3-orbits, for eighteen in all. Having no better way, he divided them into three groups with six members each, which required reopening the 3_1 and 3_2 groups closed at argon. After the transition metals the fourth period parallels the third.

The payoff of this numerology comes in the sixth period. It builds out in analogy to the fifth except that in addition to opening the 5-groups to accommodate eighteen electrons, it completes the 4-groups by adding 4_4 circles and other 4-orbits. Analogy suggested that, since the 1-group is complete with two electrons, the 2-group with two subgroups of four electrons each, and the 3-group with three subgroups of six each, the 4-subgroups should be content with eight each. Of the 32 such orbits, 18 existed in xenon, leaving 14 for the very similar elements constituting the rare earths. According to Bohr's count, the last of them should be element 71. Mendeleev had left a blank at 72 and most chemists supposed that it was a very rare earth. Moseley had examined some candidates with his X-ray techniques in 1914, but none had given off the tell-tale lines. Bohr concluded that rare-earth chemists would never find 72; it was not a rare earth, but, as the table of elements suggested, a homologue of zirconium.

Perhaps no one but Bohr could have impressed his puzzled audience with this inspired hand waving. And perhaps only Bohr could have arrived at one of his truths in this way. Using improvements on Moseley's methods, Hevesy and a Dutch spectroscopist at the Institute, Dirk Coster, immediately found 72 in all the specimens of zirconium they examined. The missing element was not rare at all. Although some French scientists claimed to have found lines of 72 in a rare-earth preparation, and complained that the 'Danes' were again trying to bag the profits as they had done during the war, the evidence was irresistible. The 'Danes' (a Hungarian and a Dutchman!) therefore had the right to name 72. They named it hafnium after the Latin for Copenhagen. It became a word of art at the Institute, 'hafnium content' signifying the amount of truth in the increasingly crazy attempts to solve the quantum riddle.

The Bohrfest in Göttingen was followed a year later by a Bohr Heft, a special issue on Bohr's theories in the German equivalent to the British journal *Nature, Die Naturwissenschaften*. Between these appreciations from Germany, the neighbourly Swedes acknowledged the importance of Bohr's work with the Nobel Prize for Physics for 1922, when they also awarded Einstein the prize for which they had not found a worthy recipient in 1921. Bohr dramatized the printed version of his Nobel lecture with the up-to-the-minute news that Hevesy and Coster had discovered the element he had predicted precisely where he had told them to look for it.

In a short speech at the Nobel banquet Bohr called attention to the role that he and his Institute were playing in promoting international science. His own work, he said, had 'consisted in joining together contributions to our knowledge of nature which we owe to investigators of various nations, who have built on widely different scientific traditions'. He had in mind his connecting Thomson's and Rutherford's experimental investigations of atomic structure with the theoretical speculations of Planck and Einstein. The dividend from linking research originating under 'different human conditions' was prized, and often reaped, at the Institute.

Bohr stressed his 'undeserved good fortune' in being able to serve as a connecting link. He was well aware that he had multiplied the pile of debt he had mentioned so often to Margrethe before their marriage. Now there was the war debt—'how undeserved[ly] favoured I feel to be able to spend my time on scientific work during this sad state of affairs in almost all the rest of the world'—and the Institute. He had to pay back. And so, with great regret, he refused an offer from the Royal Society of London of a personal research professorship tenable at Cambridge. It was very tempting. He would have privileged access to the results of the great research school that Rutherford, who had succeeded Thomson, was building there. But against the Cantabridgian truth that Bohr and Rutherford together would make a mighty force against nature stood the Hafnian truth that in Denmark Bohr could mobilize a pan-European attack on the same objective. Bohr tried to realize the potential of both truths. Could he not split his time between professing in Cambridge and directing in Copenhagen? The Royal Society, knowing that the opposite of many truths is clarity, said, 'no.'

15. Heisenberg, Pauli, and Bohr in the Institute's lunchroom.

Among the attendees at the Göttingen Bohrfest were Sommerfeld, Heisenberg, and Heisenberg's great friend and former fellow student Wolfgang Pauli (Figure 15). Though two more different types would be hard to find—Heisenberg was a clean-cut, clean living, nationalistic boy scout, and Pauli a night owl given to the cinema, cabaret, and other bad habits—together they managed to solve the two great problems that had worried Bohr for a decade. Pauli came to Copenhagen for a semester immediately after the Bohrfest. He would solve the problem of the periodic structure of the elements with an Exclusion Principle undreamt of in anyone's philosophy. Heisenberg had to postpone his pilgrimage until he finished his studies under Born. From September 1924 to May 1925 he was in Copenhagen on a Rockefeller Foundation fellowship. With Kramers's help he quickly became adept in CP magic, and in May 1926 he replaced Kramers, who had returned to Holland, as Bohr's assistant and co-worker. Their fiery and fruitful collaboration, which produced the Uncertainty Principle and Complementarity, lasted until Heisenberg received a call to the University of Leipzig in 1927.

Exclusion and energy

Pauli experienced the shock of his life in Copenhagen: he took up a problem he could not solve. The problem centred on the complicated version of the Zeeman effect suffered by spectral lines in a weak magnetic field. Explanation of these complex patterns was made additionally tormenting and tantalizing by their coalescence under a strong field into the simple triplet discovered by Zeeman. As a crutch to interpretation, Pauli and Bohr had to hand a quantum number *j* introduced by Sommerfeld in 1920 as a way to classify the energy sublevels implied by the existence of complex spectra. Bohr had not made use of *j* in distributing electron orbits because he could not find an 'unambiguous' way to introduce it via the CP. As he wrote to Coster in December 1924, 'we do not yet have any possibility of connecting the classification of levels in a rational manner with a quantum theoretical analysis of electron orbits'. Or, more oracularly, 'the difficulties with which we are fighting are the want of an unambiguous basis for the classification of the levels under consideration by means of quantum theory symbols related to the electron orbits'.

While Bohr and Pauli sought valiantly but vainly for an unambiguous interpretation of their symbols, Born and Heisenberg proved that no reasonable mechanical model could be manipulated to give the energy levels in normal helium. Bohr did not regard their finding, as they did, as a catastrophe. It just demonstrated what he already knew: further progress required further innovation. He expected it to come from Pauli. But Pauli failed: he had to content himself with classifying the lines of the anomalous Zeeman effect and tracing them through their amalgamation into the normal triplet. He considered the result 'abominable' (a word rapidly becoming a term of art in atomic physics), as he had to use two quantum numbers unrelated to orbits to succeed. 'I am very depressed [he wrote] that I have not been able to find a satisfactory explanation of these dumbfoundingly simple regularities in terms of a model.' Heisenberg delighted in Pauli's failure. '[N]o one understands quantum theory any more. That I find very agreeable... Born describes our task for the near future as "the discrediting of atomic physics."' Pauli fled to a strictly classical problem. 'It was very good for me to withdraw from atomic physics for a while [and from] problems too difficult for me.' A new try by Heisenberg at the anomalous Zeeman effect disgusted him; 'purely formal', he objected, 'devoid of physical ideas', 'ugly', 'insulting', 'unphilosophical'.

Criticism of Heisenberg's unphilosophical approach returned Pauli to atomic physics and a powerful discovery: the CP was not the only way forward. He ascribed the complications that demanded the introduction of j not to concealed properties of the orbits, but to the duplicity of the orbiter, to a 'classically undescribable two-valuedness' of the valence electron. With this transfer of sin he could exploit a classification of electron orbits made by Rutherford's student Edmund Stoner, who had suggested that the number of electrons in a given quantum state n_{kj} is indicated by the number of complex Zeeman levels. Following up Stoner's hint and applying the four quantum numbers he had used previously (one of them now supplying the electron's duplicity), Pauli declared that the reason for the closure of the electron shells, the reason for the periodic properties of the elements, is the impossibility of fitting more than one electron with a given set of the four quantum numbers into the same atom. Since theory allowed n values of k for a given n and $2k - 1$ values of j for each k, Pauli had n^2 possibilities for every n; which, when doubled for electronic duplicity, gave $2n^2$ as the maximum population of shell n. These were just the numbers wanted: 2, 8, 18, 32. Moreover, Pauli's scheme kept once closed subshells permanently closed, in accordance with the principle of the persistence of quantum numbers. Spectroscopists quickly confirmed Pauli's distribution in both the optical and X-ray regions.

Pauli's Exclusion Principle had the downside, as he wrote at the end of the paper announcing it, of escaping CP certification. Indeed, its peculiar form, which did not, could not, specify a force to effect the exclusion it demanded, seemed to him to be incompatible with ordinary systems like electron orbits. Pauli therefore opposed as counter-revolutionary the suggestion made by several physicists that the electron's duplicity arose from its spinning around an axis through its centre. 'I believe that energy and momentum values of the stationary states are something much more real than the orbits. The (still unattained) goal must be to deduce these and all other physically real, observable characteristics of the stationary states from the (integral) quantum numbers and quantum theoretical laws.' Pauli's paradoxical assignment of greater reality to properties of moving electrons than to the motions themselves was soon justified.

As Bohr's associates came to recognize that their model of multiple-periodic orbits could not account quantitatively for the energy levels even of normal helium, they accepted what he had preached from the beginning, that the orbits were false gospel. They were only a means of giving purchase to the CP and, taken in pairs, of representing quantum discontinuity. By 1920, he had almost given up on his own programme of advance. He then contemplated the desperate act of surrendering the conservation of energy, which everyone assumed to hold in the stationary states and quantum transitions. Around New Year 1924 he was negotiating terms of this surrender on the basis of a model from the New World.

Bohr spent two months late in 1923 in the USA visiting physics laboratories and money factories. From the latter, the Rockefeller philanthropies, he secured approval of an application for $40,000 to enlarge his Institute. From the former, he received an indication of the latent strength of American physics and several job offers, one at four times his Danish salary. He declined all of them because (he said) of his debt to Denmark. On the intellectual side he had the satisfaction of participating in heated debates over Arthur Compton's discovery that X-rays change frequency when scattered from electrons. Bohr opposed Compton's explanation of the effect as a collision between a light particle (Einstein's light quantum) and an electron. He still clung to the classical understanding of radiation as an electromagnetic wave, and to the relegation of quantum mysteries to the interaction of aether and matter. Moreover, he could not overcome his objection that a light quantum was a contradiction in terms: its definition employed the concept of frequency, which could be defined and measured only via the wave it was intended to replace.

Other acquisitions from the USA included students, who in their totality made up the largest foreign national contingent at the Institute before 1940. One of them, John Slater, was waiting for Bohr when he returned from his American tour of 1923. Slater brought with him a theory that gave a quasi-reality to the radiation from the set of resonators Kramers had used in calculating the intensities of the overtones emitted by orbiting electrons: quasi-real, because, in Slater's picture, the resonators' radiation carried no energy. Its role was to induce atoms it enveloped to change their state. In effect, Slater replaced the orbit described in a stationary state S by two sets of Planck resonators with frequencies corresponding to states reachable from S: one set, finite in number, for emission, another, of infinite extent, for absorption. The quasi-real or 'virtual' radiation from these virtual oscillators (physics is often fiction) gave rise to spontaneous emission in its parent atom and stimulated emission and absorption in distant atoms. Exchange of energy occurred through exchange of light particles guided by the virtual radiation.

Bohr and Kramers liked Slater's idea once they had freed it from the obnoxious light particle. They made the virtual radiation induce probabilities, measured by the Einstein coefficients, for spontaneous and stimulated emission, and for absorption, and left energy to conserve itself statistically. (Since changes in stationary states could not be correlated, neither energy nor momentum could be conserved in individual performances.) To Bohr the scheme had the decisive advantages of retaining a continuous space-time description for radiation and its interaction with matter in cases like dispersion, where no quantum jumps occurred. The scheme had a fatal flaw, however: it could be tested by experiment. Detailed investigation of the Compton effect showed that energy and momentum are conserved in individual exchanges between radiation and matter. Bohr took it philosophically. 'There is nothing else to do than to give our revolutionary efforts as honourable a funeral as possible.' Nature had spoken: Bohr would obey, and take the light quantum, or photon, seriously.

Although Slater allowed himself to be persuaded by Bohr and Kramers that their modifications of his idea improved it, he felt that he had been bullied into acquiescence. A nice letter from Bohr apologizing for not having followed Slater's original version did not dispel his resentment. There was more in this than a difference in personality. It pointed to a culture clash that would remove Bohr from the van of microphysics. Slater had gone to Copenhagen expecting that he would learn the mathematics beneath the qualitative and imprecise arguments in Bohr's papers. 'The thing that I convinced myself of after a month', he recalled, some years after the month, 'was that there was nothing underneath.' He described Bohr as a lazy mystic, trying to take the strongholds of nature by speculation rather than calculation. Slater regarded himself as a matter-of-fact person, interested in solving particular problems, pragmatic, resourceful; in a word, an American type, in healthy contrast with 'the magical or hand-waving type, who, like a magician, waves his hands as if he were drawing a rabbit out of a hat, and is not satisfied unless he can mystify his readers or hearers'. As Americans became competitive with Europeans in theoretical physics, Bohr's style of philosophical enquiry, which he passed on to Pauli and Heisenberg, and which, in a modified way, Einstein shared, would expire. Meanwhile it had productive work before it.

Chapter 4
Enthusiastic resignation

Quantum talk

Bohr's struggles to find the right word were as desperate as a poet's. And, because he wrote or revised in three languages, he had the additional problem of finding equivalents in the others of a word he deemed satisfactory in one. With the Bohr–Kramers–Slater theory, the programme for advance in quantum physics became explicitly an exercise in language building. The analogues were already in hand for frequency and intensity. A guide to inventing syntax also existed, in the CP for those who could use it, and one for increasing vocabulary, in the criterion, increasingly stressed, that the terms of the new language should refer only to directly observable quantities.

The first major contributor to the syntax of quantum mechanics was the only man fluent in what Einstein called 'Bohrish' other than Bohr himself. This was the polyglot Kramers, who shortly before Slater arrived in Copenhagen had tackled a problem that resembled the one that had seduced Bohr into atom building in 1912. But whereas Bohr had had to abandon his investigation of the response of atomic electrons to passing alpha particles because ordinary physics tore atoms apart, Kramers could employ the developed Bohr model to explore the core aether–matter question, how the electronic structure reacted to light. He began where he had in his analysis of the intensity of spectral lines, with the classical representation of the electronic motion as the vibrations of a collection of resonators.

When stimulated by a light wave, these resonators send out secondary waves at the same frequency as the primary. An electron in a Bohr atom, however, especially one at the correspondence limit, has many more opportunities than absorbing and regurgitating radiation at the frequency of the incoming wave. It might be pushed by the radiation (the laser effect) or jump spontaneously to a lower energy; processes that, as they work oppositely to absorption, had to be subtracted from the classical expression to suit the CP. The subtraction brought Kramers's formula for the dispersion and scattering of light into the same general form in the CP limit as the exemplary case of frequencies: a quantity capable of continuous variation (like $\tau\omega$) becomes asymptotically equal to a quantity available only as a finite difference (like $\nu(n+\tau, n)$). The language of continuity in classical physics is the differential calculus. The quest on which Kramers embarked, on his own and in collaboration with Heisenberg, and with Bohr looking over their shoulders, was to discover a systematic way to transform the calculus of differentials into one of differences applicable to the atomic domain.

Kramers emphasized that the new calculus should work only on observable quantities ('such quantities as allow of a direct physical interpretation on the basis of fundamental postulates of the quantum theory of spectra and atomic constitution', in Bohrish); do not look to it for any 'further reminiscence of the mathematical theory of multiply periodic systems'. There was to be no trace of the scaffold when the building was completed. How then should physicists regard the disposable electronic orbits and virtual resonators still in play? '[It] is meant only as a terminology suitable to characterize certain main features of the connexion between the description of optical phenomena and the theoretical interpretation of spectra.' It is a way of speaking the truth, but not the whole truth.

Heisenberg's famous breakthrough in the summer of 1925 finished the antechamber of the new building and dismantled the scaffold. Using the formal analogues $a(n+\tau, n)$ and $\nu(n+\tau, n)$ to $A_\tau(n)$ and $\tau\omega_n$, and replacing differentials with differences, he rewrote Sommerfeld's quantum condition on the angular momentum in terms only of quantities immediately deducible from measurements. Despite its opacity, Heisenberg's expression for h as a difference between absorption and emission terms together with the classical equation of motion provided enough information to solve for the as in terms of the νs, ns, and τs. Heisenberg illustrated the technique for the case of an imperfect (anharmonic) resonator, but could not manage the hydrogen atom. No more could Born or his assistant Pascual Jordan, who together transformed Heisenberg's intuitions into the formal mathematical language of matrices. Pauli succeeded with hydrogen late in 1925. The equation of motion they all employed was Newton's second law expressed in terms of matrices rather than classical quantities. The treatment thus implied that quantum mechanics was a theory of particles.

By the autumn of 1925 Bohr's project of discovery seemed happily fulfilled (Figure 16). His favourite student and disciple had found a basis for a coherent description of the phenomena to which atomic electrons give rise. The mathematical language required already existed in matrices, which, as square arrays of numbers, proved to be natural representations of the states available to an atom. For, as we know, it takes two principal quantum numbers to define all possible states, n and τ, and thus a plane rather than a line to write them all down. It was a most remarkable accomplishment, achieved, almost by Bohr's willing it, by ingenious young sprouts matured in his intellectual hothouse. The dozen years of ups-and-downs, of wrestling with the riddle, had been worth it. There was great joy in Copenhagen. 'Due to

16. Bohr as he appeared around 1925.

the last work of Heisenberg prospects at a stroke have been realized which although very vague[ly] grasped here have for a long time been at the center of our wishes.' Thus Bohr wrote to Rutherford on 27 January 1926. But in that same month came the news that the agonizing project that had led to matrix mechanics might not have been necessary. Physicists who had not been to Copenhagen had found an alternative atomic mechanics easier to use and free from the puzzles of the CP. It would have depressed a less resilient thinker than Bohr.

The challenger

The alternative originated in the doctoral thesis of an outsider, who never thought to leave France and could scarcely be persuaded to go to Stockholm to collect his Nobel Prize. This was Louis de Broglie, then working on his doctoral thesis, for which he invented a ghostly undulation resembling Slater's virtual radiation to regulate the behaviour of atomic electrons. He thought it reasonable that electrons in stationary states should satisfy the condition that their orbits contain an integral number n of wavelengths λ of the ghost wave. Geometry then required $n = 2\pi a_n/\lambda$. To agree with the Bohr–Sommerfeld condition, $p_n = nh/2\pi$, de Broglie had to take $\lambda_n = h/p_n$, which connected material particles with a wave in the same manner that light quanta related to radiation. The business perplexed de Broglie's examiners. They applied to Einstein, who again contributed essentially to quantum physics by alerting the philandering professor of physics at the University of Zürich, Erwin Schrödinger, to de Broglie's waves. Schrödinger had played briefly with Bohr orbits (he had demonstrated that the valence electron of the alkalis had to spend much of its time within the core of inner electrons), but had not played for long; he was a faithful product of the Vienna school founded by Ludwig Boltzmann, a mastermind of late classical physics. Schrödinger gave the French ghost wave a Viennese body by making it satisfy a classical wave equation.

Like any confined wave, Schrödinger's can be stationary, with various numbers of dead spots or nodes; we are back to plucked guitar strings, or, better, struck drumheads. Since Schrödinger's Ψ wave exists in three dimensions, its stationary solutions have three sets of nodes counted by three sets of integers. Schrödinger took these integers to be quantum numbers! Let them be n, k, and j. Each triplet of numbers defined a stationary state and the Ψ-equation enabled calculation of the state's energy E_{nkj}. All Schrödinger needed to complete his theory was an interpretation of Ψ. One lay ready to hand in his treatment of hydrogen: let $e\Psi^2(x,t)$ represent the density of electric charge at the point x at time t. A cloud of electric charge thus constituted a stationary state, and passage from one vibratory mode to another made a radiative process. It was cloudy, perhaps, but classical, as were the calculations, which used standard mathematics to calculate energy. It took Pauli months to obtain the hydrogen levels with Heisenberg's matrices, Schrödinger weeks or maybe only days to do the same with his Ψ wave.

Heisenberg did not have Bohr's resilience. He regarded Schrödinger's approach as a counter-revolutionary putsch, an attempt to return the quantum grail he had just grasped to the shrine of classical physics. It was 'revolting', even if easier to use; but calculation is just technology, after all, and the Ψ wave itself just 'trash' (*Mist*). Schrödinger reciprocated these compliments: the discontinuities in quantum physics were 'monstrous... almost inconceivable', and matrix mechanics 'repelle[nt]'. He had the satisfaction of having Einstein on his side. '[T]he idea of your work springs from true genius... I am convinced that you have made a decisive advance with your formulation of the quantum condition, just as I am convinced that the Heisenberg–Born method is misleading.' During the summer of 1926 Born added to the Göttingen method an interpretation of the Ψ wave based on the sort of experiments that had led to Rutherford's atom. It became standard: $\Psi_{rs}^{\ 2}$ measures the probability that a particle originally in state r ends up in state s after interacting with a target atom. That reduced the Ψ wave from a material substance (if charge density is material) to a means of calculating probabilities. Schrödinger objected. It was time for a trip to Copenhagen.

He laid out his version of the truth at the Institute on 4 October 1926. In Heisenberg's later highly coloured account of the resultant discussion, Bohr kept badgering his guest to give up hope for a return to space-time description of the atomic world. The badgering continued until Schrödinger fell ill; which, however, did not save him, as Bohr pursued him into the Institute's guest suite, and might have killed him with physics had Margrethe not intervened. Bohr's account in a contemporary letter is more plausible. 'The discussions centered themselves gradually on the problem of the physical reality of the postulates of the atomic theory.' The contestants agreed to disagree about the necessity of discontinuity. Schrödinger stuck to his conviction that stationary states with sudden transitions could be avoided. 'But I think we succeeded in convincing him that for the fulfillment of this hope he must be prepared to pay a cost.' This would be 'formidable in comparison with that hitherto contemplated by the supporters of the ideas of a continuity theory of atomic phenomena'. Bohr had himself and Kramers in mind among such supporters, and the surrender of energy conservation as only an instalment of the costs that Schrödinger would incur. Schrödinger rejected altogether Bohr's 'remarkable [belief]... that any understanding [of the microworld] in the usual sense of the word is impossible. Therefore conversation [with him] is almost immediately driven into philosophical questions, and soon you no longer know whether you really take the position he is attacking, or whether you really must attack the position he is defending.'

Schrödinger's letter of thanks for Bohr's hospitality, although diplomatically couched, suggests the passion with which Bohr pursued his version of truths, and the inspiration, despite the irritation, experienced by those he thought it worth labouring to convert. Schrödinger thanked 'the great Niels Bohr' for the privilege of 'talking with him for hours about things so close to my heart, and hear[ing] from him about the positions he now takes toward the many attempts to build a bit more on the sound foundation he has given to modern physics. That was for a physicist, who is one most earnestly, a truly everlasting experience!' They had not agreed about much. Schrödinger had acknowledged the strength of Bohr's objections, but could not accept, 'as it seems to me you do', even as a temporary 'resting place', that visualizable pictures like electron waves and orbits are only symbolic.

Nor could Schrödinger accept that his Ψ wave gave probabilities for the outcome of experiments on a great many systems and no information about the behaviour of a single one. Are there not single systems we can describe? Schrödinger: perhaps, but whatever we do we must not entertain contradictory descriptions. 'Certainly we can weaken them by saying, e.g., that the whole atom behaves itself "in certain circumstances, so, as if... and in certain [other] circumstance, so, as if...," but that is so to say only a logical quibble, which cannot be transformed into clear thought.... What I have in mind is just this one proposition: even if a hundred trials fail, we must not give up the hope of reaching the goal—not, do I insist, through classical pictures, but through logical concepts free from contradictions.' This admonition struck at Bohr's core self image. He prided himself, rightly, on his ability to detect 'ambiguities' that even an Einstein might overlook. He threw himself into ridding the new mechanics and its interpretations of contradictions.

Neither matrix nor wave mechanics solved the quantum riddle first posed by Planck's $\varepsilon = h\nu$ and, two decades later, by de Broglie's $p = h/\lambda$. Planck's form, when applied to the photo- or the Compton effect, related the energy of a particle to a frequency, and de Broglie's form, when applied to an electron beam, associated the length of a wave with the momentum of a particle. Should the priority go to the wave, which, in Born's interpretation, guides its associated particle, or, in Schrödinger's, constitutes it; or to the particle, which in Heisenberg's interpretation is the reality, the Ψ wave at best its mathematical servant?

Reconciling these apparently necessary but discrepant relations was the work Bohr was born for. He brought to the question a liking for waves because of their continuity in space (continuity being the mark of rationality in the philosophy he had learned from Høffding), and the conviction he had also held since his university days that no single truth could express the entire content of any domain of experience. Therefore, in conformity with the perfect symmetry of the expression of the riddle, $p = h/\lambda$, $\lambda = h/p$, he did not privilege waves or particles. He tried to make use of both.

Here is one way to join them. Two waves with slightly different wavelengths λ and $\lambda + \Delta\lambda$ start together in the aether 180 degrees out of phase. They therefore cancel out at $t = 0$. Some short time Δt later they again cancel. Say the combined 'wave packet' then occupies an extension Δx (Figure 17). Then the number of wavelengths λ in Δx must be one short of the number of wavelengths $\lambda - \Delta\lambda$ so that, with n much larger than 1, $(n+1)(\lambda - \Delta\lambda) = n\lambda = \Delta x$. Thus $\Delta x\Delta\lambda/\lambda^2 = 1$. Now $\Delta\lambda/\lambda^2 \approx 1/(\lambda - \Delta\lambda) - 1/\lambda = \Delta(1/\lambda)$. The quantum riddle in de Broglie's form gives $1/\lambda = p/h$ and thus $\Delta x\Delta p \approx h$; which is to say that the wave packet of extent Δx has a momentum that cannot be specified closer than $h/\Delta x$, and might serve as a representation of a washed-out particle. Since, however, the packet tends to spread, the representation is not perfect. A similar relation joins t and E.

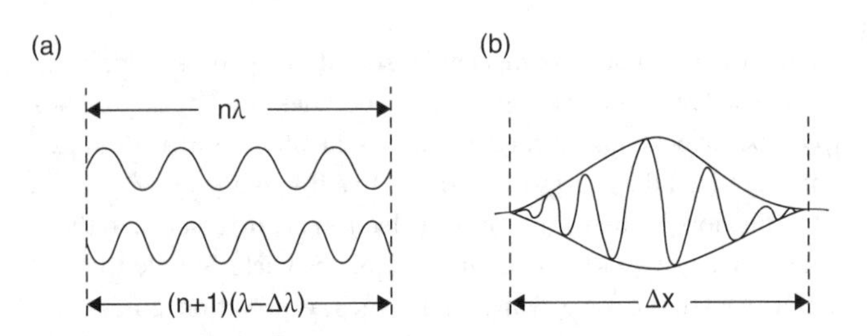

17. The wave packet: (a) a piece Δx long of trains of sinusoidal waves of wavelength λ and λ – Δλ; (b) the pieces combined into a packet.

Heisenberg derived these relations in another way in 1926, when, after intense arguments at the Institute about the riddle early in the year, Bohr went skiing alone in Norway to cool off (Figure 18). Left to himself, Heisenberg developed 'uncertainty relations'

18. Bohr cooling off in Norway.

referring to the simultaneous measurement of coordinate quantities like p, q and E, t; and, more daring yet, sent a paper about them for publication from the Institute while Bohr was away. Heisenberg may have suspected that everything was not in order in his exposition; but he knew that even if it were perfect, Bohr would insist on examining every syllable with his inexorable analysis, which on earlier occasions had driven Heisenberg to tears. He was right: Bohr found fault with the argument and made Heisenberg correct the worst error in page proof.

Complementarity

It was not for its mathematics that Bohr criticized Heisenberg's presentation, but for its one-sided description of the phenomena. Heisenberg had considered both object and instrument of experiment as particles (electrons, light quanta, atoms) and took as the quantities under investigation the particle properties, position, momentum, energy, and path. Bohr insisted on even-handedness. Notably he criticized Heisenberg's now famous thought experiment of the γ-ray microscope, through which the recoil of a light quantum reveals the position of an electron, as wrong in principle. Classical theory limits the resolution of a microscope to a separation proportional to the wavelength of the light employed (Figure 19). To locate a particle within a wavelength λ (= Δx) requires illumination by light particles with wavelength no greater than λ. But the smaller λ and the better the resolution, the greater the momentum of the photons and the more uncertain the uncontrollable (though limited) exchange of momentum between them and the observed particle. Heisenberg took as the uncertainty in the particle's momentum Δp the recoil of the photon as calculated from the Compton effect, $\Delta p = h\nu/c$, and so retrieved $\Delta x \Delta p = \lambda h\nu/c = h$.

No, no, Bohr must have ranted when he read Heisenberg's argument: if Δp was h/λ it would be known as exactly as λ, and there would be no uncertainty. That was not the greatest

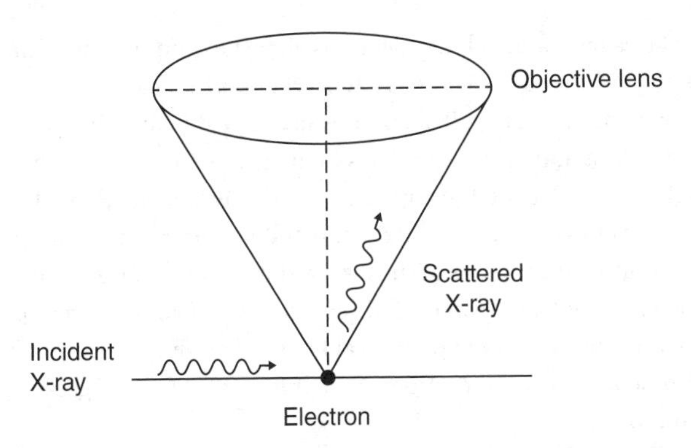

19. Heisenberg's γ-ray microscope: the electron is supposed to be located within a distance Δx of the axis of the instrument.

paralogism, however. Heisenberg had not followed through on the classical functioning of the microscope. The exact condition for the minimum resolution of an object Δx on the microscope stage by light of frequency λ is $\lambda/2\sin\alpha$, where α is the half angle of the objective lens viewed from the middle of the stage. Since the γ ray can reach the objective if scattered anywhere within the angle fflα, $\Delta p/p = 2\sin\alpha$. Thus a proper analysis makes $\Delta x \Delta p = \lambda p = h$, as before; but now there is an inescapable uncertainty in the measurement of the recoil momentum. Bohr's paradoxical method secured the uncertainty by giving the measurement instrument its full and exact classical treatment before imposing the quantum via de Broglie's rule.

Bohr laboured over his version of the meaning and lessons of quantum theory through the summer and autumn of 1927 with the help of Pauli and other regulars at the Institute. His excruciatingly leisured pace of composition, during which, like the rabbi in the story, he moved from relative clarity to sublime obscurity, accelerated toward the autumnal equinox; his commitments to present his account to his colleagues in Como (at a meeting to commemorate the centennial of Volta's death) and in Brussels (at a Solvay conference) were coming due. These conclaves had political ramifications almost as complex as Bohr's message: the small Solvay conference, supported from an endowment made by a Belgian internationalist before the war, was the first of its post-war series to invite any German but the pacifist Einstein, and the large Como meeting was the first effort of Mussolini's government to demonstrate to international science the cultural attainments of the fascist state. The conference in Brussels aimed at openness and represented the interdependence of nations in the pursuit of natural knowledge; the meeting in Como aimed at international recognition of Italy's progress in autarchy.

One thing, however, was common to both conclaves: only Bohr's initiates had the least idea what he was talking about. The most finished of the elucidations of these lectures appeared in *Nature* in April 1928. Between long periods heavy with inevitability ('it is just this very circumstance that demands...'), obligation (we must 'renounce' the classical goal of causal space-time description), and mystery ('symbolical quantum theory methods', 'symbolical utilization' of classical ideas), the persevering reader found shrewd elucidations of thought experiments. The difficulty of the presentation, apart from its language, lies in its two (or more) intertwined discourses: one, the peculiar behaviour of the microworld when the physicist tries to describe it using classical concepts; the other, association of the physicist's difficulty with deep problems of cognition, the nature of science, and the human predicament. Recognizing this double discourse helps the reader understand Bohr's exposition of complementarity.

Here is a try. The quantum postulate, symbolized by h, records the brute fact that the atomic world is characterized by 'an essential discontinuity, or rather individuality'. Consequently, any measurement made on an atomic system disturbs it in an uncontrollable way; unlike the classical case, where the disturbance is either negligible or calculable, the quantum case involves an exchange between 'measuring apparatus' and 'observed system' that cannot be determined with an accuracy greater than one quantum. The quotation marks in the preceding sentence indicate that since it is arbitrary how the experimenter allocates h between the apparatus and the system, to neither can an 'independent reality' be ascribed. It is the old problem of the observer altered by the observation.

After the measurement we know very little about the system: the concepts of a free particle and radiation in space are abstractions, things-in-themselves, which may be useful symbolically, but cannot in principle be subjects of investigation. Bohr inflated this last extrapolation from the quantum postulate into the high epistemological principle that we must 'regard the space-time coordination and the claim of causality, the union of which characterizes the classical theories, as complementary but exclusive features of the description, symbolizing the idealization of observation and description respectively'. The uncertainties in measurement destroy the ideal of observation and complete space-time coordination; the resultant imprecision prohibits the exact description of things-in-themselves and thus the exact prediction of the outcome of their interactions.

The peculiarity of the situation comes out clearly in quantum kinematics. Planck's formula $\varepsilon = h\nu$ becomes a paradox when applied, as Einstein did so successfully, to a light quantum; for its very definition requires a quantity taken from, and measurable only via, the classical theory of waves. The only way we can reconcile these ideas is to take whatever we can from the old undulatory theory, notably the concept of the wave packet, which leads to $\Delta q \Delta p \approx h$. (In keeping with later formulations, q is written for x.) The de Broglie translation rule, $p = h/\lambda$, which converts the classical description of a wave packet into a quantum description of a particle, was to Bohr a 'simple symbolical expression for the complementary nature of space-time description and the claims of causality'. This brought him to practical thought experiments illustrative of his concession that physicists *can* make observations and describe them in classical terms provided that they take the uncertainty relations into account. Bohr would later insist that his colleagues *had* to use classical concepts to describe their experiences 'unambiguously'. This extrapolation would meet significant resistance.

Matrix mechanics emerged from the concepts of stationary states and quantum transitions by 'symbolic application' of classical theories, of which it is a rational generalization. The calculus of Heisenberg, Born, and Jordan too is symbolic, and, despite its formalism, does deal largely with space and time. So too wave mechanics is but 'a symbolic transcription of the problem of motion of classical mechanics ... and [is] only to be interpreted by an explicit use of the quantum postulate'. Schrödinger's efforts at a realistic interpretation of Ψ were thus doomed; which in any case appears from the inevitable spread of wave packets and the impossibility of obtaining a Ψ of the required form for atoms with more than one electron. But there is no doubt that stationary Ψ waves make good pictures of stationary states. Bohr's representation of them by electron orbits worked because time did not enter into the description: since the hydrogen electron in a low-quantum Bohr state describes its orbit many millions of times before jumping, its space-time coordinates do not matter; Δt and Δq can be as large as the theorist required, and E and p secured to the great accuracy of spectroscopic terms. Bohr allowed that stationary states are as 'real' as 'the very idea of individual particles'; both involve 'a demand of causality [knowing E and p] complementary to space-time description' and representative of 'the restricted possibilities of definition and observation'.

Why is it that one 'elementary particle', the electron, has a charge, and the other, the light quantum, has none? Bohr suggested that a good relativistic quantum theory might explain it. Meanwhile, we must prepare for greater shocks. Relativity forced us to give up classical concepts of space and time. When wedded to quantum theory it will require more sacrifices. 'We must be prepared to meet with a renunciation as to visualization in the ordinary sense going still further than in the formulation of the quantum laws.' The great difficulty of the problem is that we lack the vocabulary for expressing its solution. The words at our disposal refer to ordinary perceptions. These we try to order and use rationally; but we have met with 'the inevitability of the feature of irrationality characterizing the quantum postulate', and can expect more instances. 'The situation... bears a deep-going analogy to the general difficulty in the formation of human ideas, inherent in the distinction between subject and object.' And with this hint at a question that had occupied him since his time with Høffding, Bohr expressed the hope, which in fact was a conviction, that the 'idea of complementarity' could deal with the irrationalities that would continue to arise as the human mind took on the universe.

Bohr's programme immediately ran into the irrational obstacle of someone else's mind. It belonged to Einstein. Every morning during the Solvay meeting of 1927 Einstein challenged Bohr with thought measurements he claimed to be more accurate than the uncertainty relations allowed. Ehrenfest was present. 'At first [Bohr was] not understood at all... then step by step defeating everybody. Speaking Bohrish ["the awful Bohrish incantation terminology"] impossible to summarize or translate.' The exchanges were like a game of chess. 'Einstein all the time with new examples... Bohr from out of philosophical smoke clouds constantly searching for the tools to crush example after example. Einstein like a jack-in-the-box: jumping out fresh every morning...His attitude to Bohr is exactly like the attitude of the defenders of absolute simultaneity towards him.'

What impressed Ehrenfest most about Bohr's arguments was the even-handed treatment of light and matter. Having secured $\Delta t \Delta E = \Delta q \Delta p \approx h$ for the light quantum via the classical argument $\Delta t \Delta \nu = \Delta x \Delta(1/\lambda) \approx 1$, Bohr turned it via the conservation laws (as in the Compton effect) to apply to matter particles. Ehrenfest again: '!!!!!!!BRAVO BOHR!!!!!!' Matrix mechanics arrived at the same uncertainty relations for particles. 'Downright undeserved magnificent harmony!!!!' Knowing then that the uncertainty relations preserve the conservation laws, the physicist can consider a collision between an electron and the moon without worrying about destroying time-hallowed principles. Still, in principle, even with the moon the concept of 'conceptual tracking...between the moments of observation' is just as false (or true) as 'tracking of a light corpuscle through the wave field between emission and absorption'. Ehrenfest ended his report where Bohr was to end his presentation in *Nature*. 'Bohr says that we have at our disposal only those words and concepts that yield such a complementary mode of description...The famous INTERNAL CONTRADICTIONS of particle theory only arise because we operate with a language that is not yet sufficiently revised.' Ehrenfest was probably right in thinking that Bohr would despair over this account of language: Bohr did not propose to alter the meaning of ordinary words in applying them to the quantum world, but to give rules for using them 'unambiguously'.

The game had two more rounds. At the Solvay conference of 1930, Einstein presented an apparently undefeatable thought measurement. Let there be a box entirely closed except for a shutter, which precise clockwork opens for an infinitesimal time Δt every quarter hour (Figure 20). Hang the box by a heavy spring so that you can weigh it with exquisite accuracy. Put a single photon in the box—how you do it is your problem. Start the clock. During one of the instantaneous openings the photon will escape. Then weigh the box to whatever accuracy you please, say ΔW. From ΔW you have ΔE ($E = mc^2$!) and so have beaten Bohr: Δt and ΔE can be made small enough that their product is less than h. After a sleepless night Bohr reminded Einstein that the theory of relativity, invoked through the equivalence of mass and energy, requires that the rate of a clock depend on its position in a gravitational field. You cannot rely on the reading of that clock, Einstein, for its rate altered uncontrollably as the box rose when

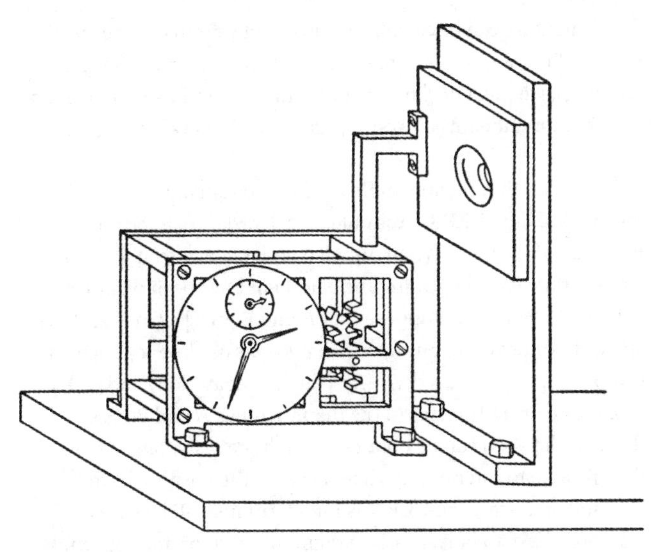

20. Einstein's photon-in-a-box experiment proposed to defeat the uncertainty principle.

the photon left; and, in fact, using your equations, the resultant imprecision is just enough to save the uncertainty relations.

Einstein tried again in 1935 with the help of two colleagues at Princeton, where he had fled to escape the Nazis. The 'EPR' argument (after Einstein–Podolsky–Rosen) did not try to defeat uncertainty but, rather, to convict quantum mechanics of not accounting for 'essential features of reality' within its domain. Quantum mechanics allows the simultaneous measurement to any degree of accuracy of the difference in position $q_A - q_B$ and the sum of the momenta $p_A + p_B$ of interacting systems A and B. Let them then fly apart to the uttermost parts of the earth. Experts awaiting the arrival of A have a choice of measuring its position or its momentum precisely; if q_A, the exact position of the now far distant B is known from the previously measured $q_A - q_B$; if p_A,

B's momentum can be calculated, also exactly, from knowledge of $p_A + p_B$. Since we cannot suppose that anything we do to A can affect remote B, we must declare quantum mechanics incomplete in not allowing an undisturbed system exactly defined values of p and q.

Bohr replied that the incompleteness lay not in quantum mechanics, but in EPR's example. For in order to determine q_A in Fermilab after A and B had mingled and escaped from CERN the circumstances of their mingling could not have been what EPR advertised. For let us suppose that A and B are electrons and the apparatus a heavy diaphragm with two slits as close together as you please. Then, indeed, $q_A - q_B$ (the slit separation) and $p_A + p_B$ (the momentum transfer to the diaphragm as A and B pass through the slits) can be obtained simultaneously; but we will not then be able to infer q_B anywhere because the position of the hanging diaphragm in CERN is undetermined. To be sure of q_B, we would have to bolt the diaphragm to the laboratory bench; but then we could not measure $p_A + p_B$. BRAVO BOHR! Einstein continued to believe that physicists could do better than quantum mechanics allowed, but he stopped trying to prove it.

In contrast to the Solvay exchanges, EPR's thrust and Bohr's parry were conducted before the large public that read the *Physical review*, the American journal then rapidly becoming the major periodical for international physics. The exchange received some attention, though most physicists who noticed it regarded it as Sabbath sermonizing of little use on workdays. Within Bohr's church, however, there was rejoicing at seeing the gospel saved by recourse to the well-known parable of slits in a diaphragm. But perhaps the most instructive responses came from C. W. Oseen, professor of physics in Uppsala, who had followed Bohr's work closely from its beginnings, and the émigré physicist-philosopher Philipp Franck. 'At last I understand', Oseen wrote, what you have been saying all along: before a measurement an atom's state with respect to the quantity measured is not defined. That was only half of it. Frank understood Bohr to mean that 'physical reality' should not be ascribed to the quantities we associate with micro-entities. Quantum mechanics as interpreted by complementarity characterized measuring procedures and results, not the things measured. Bohr acknowledged that that was what he had in mind. But that was not yet the whole story.

Chapter 5
The Institute

Mopping up

In 1927 Bohr's Institute welcomed more long-term visitors (24) and published more papers (47) than in any other year in the inter-war period. The relatively large numbers represented not only the excitement of the 'Copenhagen Spirit', as Heisenberg called Bohr's approach to quantum problems, but also American money. As we know, the Rockefeller Foundation supported expansion of the Institute. The Carlsberg Foundation and the Rask–Ørsted Foundation continued their subventions to foreign researchers, and the Danish government increased funds for salaries and maintenance. Bohr gained freedom from ordinary lecturing from the government and a large increase in salary from the Carlsberg Foundation. In 1932 he was able to add to the Institute's effective working space by moving to the villa at the centre of Carlsberg wealth production, its brewery. Bohr lived, entertained, and worked at the villa until his death in 1962 (Figure 21). Since the brewery stood some distance from the Institute, Bohr no longer could easily detain his assistants after dinner or supervise their work as closely as he had done.

The distance may have been a benefit, as Bohr told the Danish Academy when accepting its invitation to occupy the villa. Measured by visitors and published papers, the Institute's

21. Bohr and his wife in front of the Carlsberg villa.

attraction for foreign visitors had declined since 1927, reaching an all-time low in the early 1930s. One reason for the decline, Bohr supposed, was his preoccupation with theory, which had taken him into a lengthy probe of electromagnetic measurements. His collaborator in this detailed work was Léon Rosenfeld, a polyglot polymath Belgian, who would remain his assistant until war came again. Rosenfeld then fled to England and a professorship in Manchester; he returned to Copenhagen after the war to collaboration with Bohr and a professorship at the University.

There was another reason for the decline of activity at the Institute in the early 1930s. Bohr's programme and intuition blinded him to the most productive ideas in quantum physics then under discussion. He rejected, or belatedly accepted, Dirac's relativistic theory of the electron, the discoveries of the neutron and positron, the invention of the neutrino by Pauli, and the theory of beta decay founded on it by Enrico Fermi. Bohr resisted these innovations owing to commitments deeper than the old prescription against multiplying essences (in this case particles) unnecessarily. He kept looking for the limit where, according to the lesson he had learned from Høffding, quantum mechanics must meet with an irrationality in its domain that restricts its scope of application.

Although the period of continuous advance of quantum mechanics had been short, its recovery of the successes of the 'old' quantum theory and acquisition of new territory gave it a longer psychological run in Bohr's mind than in minds not as prepared as his to seek irrationalities. He had pounced on difficulties encountered in imposing relativity theory on quantum mechanics, and the conservation of energy on radioactivity, as heralds of a limit. Prematurely he announced the likely need of more renunciation, this time not of the unrestricted application of classical concepts, but of the concepts themselves. The target of his enthusiastic resignation was, again, the conservation of energy. The sacrifice had not been needed when he and Kramers forced it on Slater. No doubt that was because they had not applied it to appropriate phenomena. Beta decay, however, showed the failure of conservation clearly: the energy lost by the decaying nucleus was fixed, the kinetic energy of the escaping electron variable. Fermi's idea that beta electrons do not exist in nuclei, but are created, along with Pauli's neutrino, in (or as) the process of radioactive decay, just as photons are in atomic transitions, saved conservation. So Bohr, disappointed again, doubted the neutrino.

Surrendering energy conservation, or some other prized acquisition of classical physics, would delimit quantum physics as the theory of those aspects of the microworld that require renunciation of unrestricted application of the concepts of classical physics, but not reformulation of the concepts themselves. Between the hypothetical domain where quantum theory did not suffice (characterized perhaps by the need to take conditions for the existence of elementary particles into account) and classical physics augmented by relativity lay the territory the Copenhagen theorists had mapped. Bohr hinted to a meeting on nuclear physics organized by Fermi in Rome in 1931 what he had in mind. 'Just as we have been forced to renounce the idea of causality in the atomistic interpretation of the ordinary physical and chemical properties of matter, we may be led to further renunciations in order to account for the stability of the atomic constituents themselves.' This was to refer to posterity the ultimate challenge of explaining the astonishing stability of the material world.

For a decade after the invention of quantum mechanics, Bohr allowed the Institute's experimental programme to flounder as he tried to guide the development of relativistic quantum theory towards the irrationality he foresaw. He did not pursue new financial resources for the Institute. He continued in his comfortable way of intense, wide-ranging talk about principles and limits, and attempted to interest biologists and psychologists in the lessons he drew from quantum physics. He offered the attendees of a conference on radiation therapy a means of assuring that their domain would not follow chemistry as an annex to physics. It is logically impossible, Bohr's reassurance ran, to seek the secret of life in minute physical structure, because invasive investigation at the atomic level would kill the specimen and with it the 'life' under study. A complementarity thus exists between physico-chemical analysis and the conditions necessary for life 'just' sufficient to ensure that its description cannot be exhausted using only physico-chemical concepts. That sounded to many like vitalism. But Bohr meant only to emphasize that the concepts of physics cannot describe the apparently purposeful behaviour of living beings. He no more ascribed a life spirit to matter than he did mechanical quantities to unobserved electrons.

To psychologists he urged his comforting old lesson about freedom and determinism with the authority of the new physics. Free will and fully determined action are complementary terms applicable to different experimental situations. In general, analysing a concept and employing it stand in a relation of complementarity.

Just as the quantum physicist had to renounce classical descriptions, so other scientists and philosophers, wise enough to accept the epistemological lessons of physics, must recognize that many of the enduring problems of their disciplines existed only because their predecessors had not understood that they were not soluble, or even comprehensible, in the terms proposed. Resignation of the pursuit of Truth in favour of the admission of apparently contradictory truths in the sense of complementarity could dispose of boatloads of *Scheinprobleme*, or red herrings, in many academic backwaters.

Some of Bohr's more enthusiastic disciples declined to remain at this modest epistemological level. Pauli, whose breakdown over life and physics brought him to the consulting room of the psychologist Jung, developed during treatment a deep analogy between Jungian psychology and quantum physics, both merely ways of grasping 'the invisible and untouchable'. The entering-into-consciousness of material from the unconscious is precisely analogous to experiment or measurement in atomic physics. In both processes the analyst confronts the irrational. Jordan, who also had psychological problems but preferred to handle them in the manner of Freud, discovered a parallel between schizophrenia and the uncertainty principle. When one side, A, of a split personality subdues the other, B, the same logical situation arises as when precise measurement of position q entirely suppresses knowledge of the complementary quantity p; much as in Pauli's version the spontaneous leak from the unconscious that causes a jump from A to B is equivalent to a measurement that prefers p to q. Jordan went much further than Pauli by locating in microphysical uncertainties in the brain the causes of all sorts of mystical things: free will, telepathy, clairvoyance, Lamarckism, voluntarism. Complementarity called for these things, he said, 'one must get used to [them]'.

Most physicists shuddered at these excesses; Einstein rated them 'reprehensible'. Bohr repudiated Jordan's extravagances and worried that his doctrine might lend itself to mystical and irrational worldviews, like National Socialism. Jordan justified the worry by joining the Nazi party. Despite exploitations of complementarity in reactionary politics and subjective science, however, Bohr continued to think that, properly instilled, it could serve as a reliable guide to life and belief. According to Rosenfeld, Bohr expected that eventually complementarity would be included in general education; 'better than any religion... complementarity would afford people the guidance they needed'.

Bohr could not have expected this quantum transition in human affairs to occur soon, because, as he acknowledged to Kramers, even most physicists refused to recognize the sweeping philosophical significance of their discoveries. Why? Bohr's explanation: 'So many of the physicists, who have contributed so essentially to progress in the field, suddenly seem to be afraid of the consequences of their own work.' More probably, they agreed with Darwin that Bohr's peculiar mixture of imperialism and resignation, which they found hard enough to swallow in physics, was mere philosophy outside it.

Such weighty matters lent themselves to levity. The Institute, and Bohr too, enjoyed a little parody at his expense. The best of the play had as ringleaders George Gamow and Max Delbrück, both of whom enjoyed fellowships in Copenhagen before emigrating to the USA. Their serious sides produced, in Gamow's case, the immensely important 'tunnelling effect' whereby particles can crash through a nuclear barrier with less energy than they would require in classical physics; and, in Delbrück's case, work in molecular biology that won him a Nobel prize. Their playful sides invented ping-pong in the library with books as paddles (Gamow) and a three-page letter in one sentence as a specimen of Bohrish (Delbrück). In collaboration they produced a version of Goethe's *Faust* with Pauli as the devil, Bohr as God, Ehrenfest as Faust, and the undetectable neutrino as bashful Gretchen (Figure 22).

22. Title page of the Copenhagen *Faust* by Delbrück and Gamow, and Faust himself (Ehrenfest) in his study.

New directions

The law for the 'cleansing' of the civil service, enacted by the Nazis in 1933, expelled Jews from state offices and flooded Europe with unemployed scientists. Bohr happened to be in the USA at the time and lobbied the Rockefeller Foundation to drop its condition that its fellowship holders have jobs to return to. The Foundation decided instead to make funds available to enable senior displaced scholars to set up in new surroundings. Under this policy Bohr brought two old friends to Copenhagen, both of whom had given up their professorships although neither was forced to: Franck, who could have continued in office through a provision that

23. Franck (centre) and Hevesy at the Institute with Bohr.

excepted Jews who had fought for Germany in the First World War, and Hevesy, who had served in the war for Austria (Figure 23). Franck had quit in solidarity with Max Born and the many other Jewish physicists and mathematicians at Göttingen, and Hevesy in solidarity with his Jewish assistants at Freiburg. Their engagement in Copenhagen required money for expansion, apparatus, and salaries, and a firm research direction. The Rockefeller and Carlsberg Foundations provided most of the money, and the Rockefeller and the latest discoveries in physics pointed the direction.

The discoveries concerned the creation of new radioactive materials by sending neutrons into the nuclei of ordinary matter. By 1934, when Mme Curie's daughter Irène and her husband Frédéric Joliot discovered the effect, physicists had been transmuting atoms by bombarding them with fast charged particles for several years. These projectiles came either from natural sources or, increasingly, from apparatus that began operation in 1932: a high-tension machine developed in Rutherford's laboratory by John Cockcroft and Ernest Walton, and the cyclotron invented by Ernest Lawrence and his associates at the University of California at Berkeley. Everyone assumed that the transmutations thus provoked occurred instantly, the incoming particle knocking out a nuclear one as in a game of marbles. The Joliot–Curies found that in several cases the transmuted nucleus could remain excited for some time, even days or weeks. Using neutrons from a natural source (radium-beryllium), Fermi and his group in Rome ran through the periodic table as quickly and as fully as they could, making the many new activities that would earn Fermi the Nobel physics prize for 1938. The Cockroft–Walton machine and Lawrence's cyclotron increased the yield of 'induced activities' by bombardment with protons, deuterons (nuclei of heavy hydrogen, first detected in 1932), and alpha particles, and also won Nobel prizes.

Some of these fleeting radioactivities lent themselves to an extension of the tracer method that Hevesy had pioneered in 1913 to label reactants and follow them through chemical processes. After 1934 an ever-greater supply and variety of man-made activities, especially radio-phosphorus manufactured in the cyclotron, brought life processes within reach of Hevesy's method. The Rockefeller Foundation had made the improvement of

24. Bohr amid the high-tension apparatus at the Institute in 1938.

biological work by the use of physical methods a priority. Bohr successfully applied to it for funds for experimental biology and the large apparatus needed to supply radioisotopes for tracer work and, it was hoped, for cancer therapy (Figure 24). The equipment also could be used by Franck to study the nucleus. The Rockefeller Foundation understood the connection, and approved; it supported several such accelerator labs, notably Lawrence's, which did produce something useful in therapy (radio-iodine for thyroid problems) and a fine organic tracer (radio-carbon). But there was always high tension between the biological and physical applications of the machines, especially when experimental animals turned up in the labs, as they eventually did at the Institute.

Hevesy stayed in Copenhagen building up and enlarging his programme in collaboration with several institutes for medical and biological research in the city. The programme succeeded especially in explorations of metabolism, which earned Hevesy the Nobel prize in chemistry for 1943. Franck did not stay on. Although when in Göttingen he had declined to accept any atomic theory until certified by Copenhagen, he had governed his own institute and could follow his research bent. In Copenhagen he feared that Bohr's overwhelming personality and originality would suppress his freedom of thought and action. After a year or so with Bohr he took a post in the USA, where he was to contribute significantly to building the atomic bomb.

Franck's effective replacement in Copenhagen was a man much his junior, Robert Otto Frisch, an Austrian whom Bohr supported on fellowships until the war. During Frisch's regime the Cockcroft–Walton machine began to operate and, in 1938, the cyclotron. This last accomplishment required fine-tuning by a man from Berkeley, and, while visiting Lawrence's laboratory during a round-the-world trip in 1937, Bohr had asked Lawrence to send one to Copenhagen. The only other centre in Western Europe (excluding Britain and the Soviet Union) with a fully operating cyclotron before the Second World War was Joliot's laboratory in Paris, also built with money from the Rockefeller Foundation and a helping hand from Berkeley.

Bohr now looked inside the nucleus and found there a liquid drop consisting of protons and neutrons sloshed together. He was thus in a position to supply an explanation for the conclusion reached by Frisch and Frisch's aunt, Lise Meitner, in December 1938. Meitner, a Jew, had then only recently and narrowly escaped from Berlin where, protected by her Austrian citizenship and Max Planck, she had managed to continue her work at the Chemical Institute of the Kaiser-Wilhelm Society, carried out with Otto Hahn and Fritz Strassmann, on identifying the transmutation products of uranium irradiated by neutrons. (Lacking a cyclotron, the team used desktop apparatus and a radon-beryllium source.) Following general expectations—one that Fermi had then just exploited in his Nobel-prize address—the group had supposed that in bombarding uranium they had produced 'transuranic' elements. But after Meitner's flight, Hahn and Strassmann had decided that the suppositious transuranics had the same chemistry as certain elements around the middle of the periodic table. They informed Meitner, who told Frisch; and together aunt and nephew worked out that the liquid drop constituting the uranium nucleus could split after swallowing a neutron through a process they likened to cellular fission—an echo of the programme in experimental biology forced on physicists by the Rockefeller Foundation.

Bohr worked out that fission occurred in susceptible nuclei when the captured neutron caused unstable oscillations of the liquid drop. He urged Frisch and Meitner to publish their findings immediately and set sail with Rosenfeld for a long-planned trip to the USA. Not knowing that Bohr had promised not to advertise fission before its discoverers had published it, Rosenfeld disclosed it. Confirmation came swiftly from several American laboratories. Joliot and others soon showed that a susceptible nucleus released more than one neutron per fission, making possible a chain reaction; while Bohr and other theorists realized that only uranium atoms constituting the rare isotope of atomic weight 235 (^{235}U) were susceptible. Meitner had been more right than wrong: most of the neutrons absorbed did produce transuranic elements, by converting the heavier plentiful isotope ^{238}U into a short-lived activity, which decayed into a long-lived one. The men who later separated these activities at the Berkeley cyclotron named them neptunium and plutonium, respectively, after the names of the planets beyond Uranus. ^{239}Pu turned out to be fissionable.

As physicists came to contemplate chain reactions in fissionable isotopes, the world again fell to war. Scientists on both sides understood the menace and opportunity of atomic weapons. Among the several German institutions that looked into the matter, the then new Institute for Physics of the Kaiser-Wilhelm Society predominated. Several dark ironies surrounded the venue. The Rockefeller Foundation had offered to pay for the building and apparatus at Planck's urging well before the Nazis came to power and felt bound to honour its commitment whenever the German authorities called on it. And so a funding programme set up to encourage international scientific cooperation and a physicist respected for his probity combined to give warmongers a place to develop monstrous weapons. The leader of the Institute's 'uranium project' was a one-time holder of a Rockefeller fellowship and a physicist universally celebrated for his science: Werner Heisenberg. Though neither a warmonger nor a Nazi, Heisenberg was prepared to do what he could to serve his country.

In April 1940 the Germans suddenly sailed unopposed into Copenhagen. They offered the Danish government a choice between destruction of the country and cooperation with the occupiers under the fiction that the takeover was intended as protection of a Germanic people against invasion by Britain. The government ministers agreed to cooperate on the understanding that they would run the country and that Denmark would retain its neutrality. The arrangement, though detested, worked fairly well until the summer of 1943, when German demands increased and the Danish ministers resigned en masse.

To help encourage cooperation among wavering intellectuals, the occupiers set up a Cultural Institute (as they did in other occupied countries) that sponsored lectures on the life of the mind. In September 1941 Heisenberg came to talk about astrophysics. He took the opportunity to visit his old friends at the Institute, with whom he shared his conviction that the Germans would win the war. The USA had not yet entered the fight and Germany had only to finish its conquest of Russia and settle its score with Britain. Therefore, Heisenberg urged, alignment sooner rather than later would be prudent. Bohr consented to meet Heisenberg privately. Their meeting was not cheerful. Two years later Bohr fled to Sweden with most of Denmark's Jewish population in a dramatic rescue operation at which some German officers connived. A special mission of the Royal Air Force flew Bohr to England.

Outreach

Being a man with a message, Bohr reprinted his general talks to reach a larger audience of potential proselytes than he had found among physicists and biologists. *Atomic theory and the description of nature* (1934) comprised the definitive version of his Como lecture as published in *Nature* in 1927 and three other more accessible essays initially composed between 1927 and 1929. *Atomic physics and human knowledge* (1958) has as its centrepiece Bohr's account of his discussions with Einstein; most of the other six articles concern the application of complementarity to the life sciences. Since these slim, packed volumes encapsulate Bohr's thought, it is worthwhile to review them here despite some inevitable repetition; and also appropriate because refinement through repetition was Bohr's favourite way to work.

The earliest article began as a lecture to the Scandinavian Mathematical Congress just after the invention of matrix mechanics. Adapting his message to mathematicians, Bohr emphasized the 'symbolic' character of his play with electron orbits and the deeper meaning of the numbers, atomic and quantum, specifying them. Atomic physics thus approached the

Pythagorean ideal of reducing the world to pure number. The symbolic orbital play had not been easy, as it had demanded detailed application of classical physical concepts to phenomena, like the stability of atoms, to which they were not fully applicable. The more physicists played, the more phenomena turned up, like the anomalous Zeeman effect and dispersion, which eluded exact representation by an orbital model. It appeared that some wholesome doctrine, even the conservation of energy, might have to be surrendered. In the end physicists got by with a lesser sacrifice, 'renunciation of mechanical models in space and time'.

One of the essays of 1929, an address at a celebration of Max Planck's doctoral jubilee, undertook to reassure colleagues that the renunciation of causality was not peculiar to physics, but a consequence of 'the general conditions underlying the creation of man's concepts'. These conditions related to the old problem of 'the objectivity of phenomena'. The uncontrollable uncertainty in the interaction of measurer and measured, discovered by atomic physicists, implied 'the impossibility of a strict separation of phenomena and means of observation, and [exemplified] the general limitation on man's capacity to create concepts, which have their roots in our differentiation between subject and object'. Kierkegaard knew that! Bohr again: 'No sharp separation between object and subject can be maintained, since the perceiving subject also belongs to our mental content.' If therefore meaning depends on the observer, there should be no surprise that 'a complete elucidation of one and the same object may require diverse points of view which defy a unique description'. And further: 'the conscious analysis of any concept stands in a relation of exclusion to its immediate application'. Høffding knew that!

Bohr ended this talk to Planck's well-wishers with two fanciful analogies. Item: 'The continuous outward flow of associative thinking and the preservation of the unity of the personality exhibits [!] a suggestive analogy with the relation between the wave description of the motions of material particles, governed by the superposition principle, and their indestructible individuality.' Perhaps this utterance signified that despite his discursiveness, Bohr retained unshakeable core commitments. The second analogy was the untying of the knot regarding free will with which we are familiar, amplified with the consideration that we cannot in principle find for determinism by following a causal chain through the brain. The investigation would have to be conducted at the atomic level, which would introduce quantum uncertainties and an uncontrollable psychophysical response. 'We must, therefore, be prepared to accept the fact that an attempt to observe [the causal chain] will bring about an essential alteration in the awareness of volition.' Fiddling with the brain might well affect the processes that give us the sense of willing.

The final essay in the first volume descended from a talk to Scandinavian natural scientists in 1929. It brought a change in tone. The need to sacrifice old ideas now appears in a positive light. 'Resignation' of visualizability and 'renunciation' of detailed accounts of quantum jumps record 'an essential advance in our understanding'. Classical concepts still reign within their domain, and, in the atomic world, can be used to the limit that complementarity allows. Just as we should rejoice that relativity has forced recognition of the limited applicability of our intuitions of space and time, so we should be delighted with the 'emancipation from the demand of visualizability' effected by quantum theory. Perhaps we should also be delighted by the news that, as the sages of old put it, 'we are both onlookers and actors in the great drama of existence'. Bohr ended this talk to biologists with a dictum that must have surprised them. Because of quantum uncertainty, he said, 'the distinction between the living and the dead escapes comprehension in the ordinary sense of the word'.

'Light and life', Bohr's address to the International Congress on Light [Radiation] Therapy held in Copenhagen in 1932, opens his second collection of essays. It emphasizes that photochemical processes in living beings cannot be prosecuted to the atomic level, say by following a photon and its workings into the eye and brain, for that would extinguish the functions under study. We will never be able to practise biology satisfactorily if restricted to physicochemical concepts. '[T]he very existence of life must in biology be considered an elementary fact, just as in atomic physics the existence of the quantum of action has to be taken as a basic fact that cannot be derived from ordinary physics.' This consideration provided Bohr with the leeway he required to bring physiology into agreement with psychology, in particular with the feeling of untrammelled volition. Bohr closed his talk by denying its implications of vitalism and spiritualism.

His next attempt on biology that he thought worth reprinting resulted from an invitation to help celebrate the 200th birthday of Luigi Galvani in 1937. Bohr prefaced his usual summary of quantum physics and its lessons with a brief account of classical physics from Galileo to Planck via Galvani. Then he approached physiology via psychology, saved free will while repudiating mysticism, and arrived at the limit of application of physical concepts in the description of life. The conclusion that life must be admitted as a 'basic postulate of biology, not susceptible of further analysis' supported, he said, the interpretation of 'biological regularities as representing laws of nature complementary to those appropriate to the account of the properties of inanimate bodies'. Thus a new complementarity: description of living beings/description of mechanical things.

The next four essays in *Atomic physics and human knowledge* deal with applications of complementarity to anthropology, Einstein, and human knowledge. To the International Congress of Anthropological and Ethnographical Sciences held in Copenhagen in 1938 he presented the unconvincing complementary pairs thought/feeling and reason/intellect, and the unlikely proposition, widely interpreted as a criticism of Nazi ideology, that the 'gradual removal of prejudice is the common aim of all science'.

With Bohr's defeat of Einstein we are familiar. The essays on knowledge, of 1954 and 1955, summarize the general epistemological position as Bohr had refined and rehearsed it over two decades, but do not bring much new.

The lecture to the Danish Medical Society has the unpromising title, 'Physical science and the problem of life', and, indeed, it does not go beyond 'Light and life' in substance. Psychologically, however, it represents redemption. It features a long extract from a paper of 1910 on pulmonary problems in which Bohr's father wrestled with the concept of purposiveness in physiological functions. Christian Bohr decided that although the assumption of purposiveness is a natural, useful, and even indispensable heuristic device, it must give way if physicochemical investigation demonstrates in detail how the purpose is achieved. Bohr's epistemology cleared up the matter. Complementarity allows, indeed may require, that finalistic arguments, insofar as they express the unanalysable concept of life, figure in physiologists' description of the phenomena in their domain. Thus atomic physics found a place for teleology in biology, and Niels Bohr paid back an instalment on the debt he fancied he owed his father.

Bohr too was a father, of six sons, for one of whom he contracted a most terrible debt. This was the eldest, called Christian after his grandfather. In 1934, at the age of 17, having just finished his school examinations, Christian drowned in a sailing expedition captained by his father. Six weeks later, when his body washed up in Sweden, the Bohrs held a memorial service at which they announced that Christian's great-aunt Hanna had endowed an annual grant in his name to support a young artist; for Christian, though successful in all his studies, was beginning to identify with artistic rather than scientific endeavours (Figure 25). Bohr's talk on this occasion, and on the award of the first grant in December 1934, was not halting or complicated. He could be eloquent when moved, and often recited poetry, of which he knew much by heart, with perfect emphasis and clarity. In his farewell to his son,

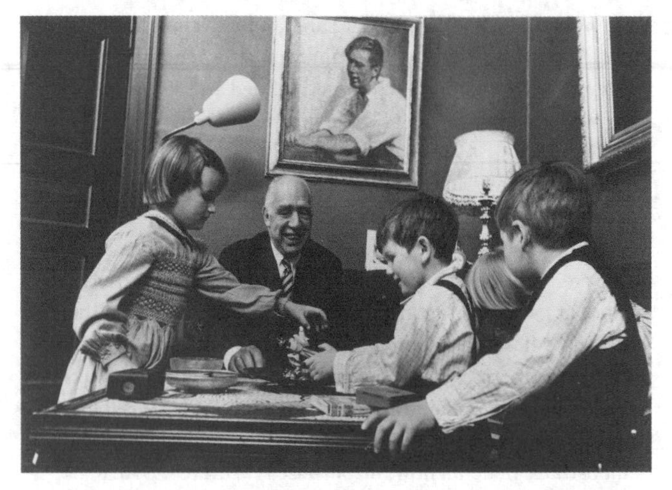

25. Bohr playing with his grandchildren in the salon of the Carlsberg villa under a portrait of his son Christian.

Bohr must have had Kierkegaard in mind, for he praised Christian above all for the same quality that he had admired in the romantic philosopher: 'the conviction that only honest work to clarify for oneself every large and small question, no matter how remote this might seem from the demands of daily life, is our one way to get a feeling for a deeper harmony behind the vicissitudes of existence'.

Chapter 6
Elder statesman

Uncle Nick

Soon after Bohr's arrival in England in 1943, the leaders of the British uranium project, which operated under the code name Tube Alloys, informed him in strict confidence of the likelihood that the Americans would complete an atomic weapon within a year. There were still difficulties, for example, how to detonate a plutonium bomb, on which Bohr's opinion might be helpful. The leader of the Manhattan Engineering District, to give the American project its full code name, General Leslie Groves, allowed Bohr access to secret material and must have done the same for Bohr's physicist son Aage, then 21, who would be the sounding board Bohr always needed when working out his ideas. At Los Alamos, which he visited several times as 'Nicholas Baker', he advised about the general prospects of the Project. But 'Uncle Nick', as physicists who liked to joke with security called him, soon took as his major assignment a committed effort to ensure that the lesson of the atomic bomb would be used for the benefit of humankind.

He reasoned in his familiar fashion that with atomic weaponry humankind faced a situation from which it could exit to its continued improvement only by renouncing views previously thought irrefragable. Just as the quantum required some sacrifice of visualization, so the atom bomb demanded some surrender of national sovereignty. If the USA and the UK, which happened to be ahead in exploiting atomic energy, were to disclose their know-how in return for an agreement among participating states to submit to international control under safeguards firmly imposed by the nascent United Nations, everyone's security would be ensured; instead of wasting their wealth in an arms race, the peoples of the world would be able to mingle their various viewpoints, their versions of the truth, to universal advantage. Bohr sometimes added a bit of the complementarity he had presented to anthropologists in 1938: all cultures are precious since their different approaches to knowledge, though often incompatible, together exhaust the sum of human experience. The experience of atomic physicists was particularly apt. They had shown the way to mingle viewpoints in the cooperative international effort spearheaded in Copenhagen that had resulted in the discoveries behind the bomb. Atomic physicists thus had both the obligation and the expertise to lead the human race out of the grave danger into which they had propelled it.

Bohr discussed these ideas in the spring of 1944 with J. Robert Oppenheimer, who had been a fellow in Copenhagen, and others able to share Manhattan secrets. He went on to bring them to the attention of the politicians who would decide the post-war fate of the world. Despite having no brief or official appointment by any government, Bohr managed by force of personality and reputation to deliver his message verbally to Roosevelt and Churchill. His route to the Prime Minister ran through the British ambassador to the USA, Lord Halifax, and the head of the British bomb project, Sir John Anderson, both of whom saw merit in Bohr's approach. Churchill did not, however, and at one point wanted to lock up Uncle Nick as a security risk. Bohr at first had better luck with Roosevelt, whom he reached through a pre-war acquaintance, Felix Frankfurter, a justice of the US Supreme Court and close friend of the President. Frankfurter and Roosevelt shared Bohr's fear that an arms race in atomic weapons would occur after the war, and his hope that the magnitude of the menace might induce the parties to sacrifice sovereignty to security. Roosevelt and Churchill did not see, however, how to sell the public (or themselves) on renouncing an advantage, however temporary, in nuclear capability. At their meeting in Quebec in September 1944, they formally rejected Bohr's approach.

When Bohr returned to a hero's welcome in Denmark and the world could see the horror of atomic bombs, he republished the views he had urged in secret and, when allowed, in the *New York Times* in the hope that the wider public and the United Nations would be more responsive to them than the leaders of the free world. The UN established an Atomic Energy Commission, which, during its two years of existence, considered a proposal by the American representative, Bernard Baruch, for some sharing of information and control. The Soviet Union naturally wanted its own demonstrated capability, not a description of American technology controlled by international authority, and its representative rejected Baruch's plan. Bohr then switched his emphasis to advocating an 'open world' in which all countries would make all information about their strategic plans and weaponry available to inspectors, thus simultaneously securing both control of existing threats and the means for collaboration in the multiplication of knowledge. He proposed his open world in an open letter to the United Nations in 1950. It was not a propitious time. In 1949 the Soviet Union had exploded its first atomic weapon. In 1950 the Korean War and the hydrogen bomb loomed, espionage was rampant, and neither superpower could find a good reason to trust the other.

President Eisenhower's unexpected proposal for sharing some technical know-how and strategic materials as an earnest of 'Atoms for Peace' gave Bohr some reason to think that his approach might eventually prevail. Denmark set up an Atomic Energy Commission with Bohr as chairman in 1955 to take advantage of the new American openness, and the Thrige

Foundation, which had given him the magnet for his cyclotron in 1938, supported an initiative at the Institute to inform engineers about atomic energy. Since many countries participated in this Cold War thaw, Bohr saw materializing the international cooperation among scientists that he hoped would guide the way to an open world. The Americans recognized him as the grand cicerone of this direction of travel. In 1954 Columbia University awarded that 'Distinguished son of Denmark [and] benefactor of the human race', Uncle Nick, an honorary doctorate. In 1957 Bohr received a greater accolade, the Atoms for Peace Award, presented by President Eisenhower in acknowledgement of 'the importance to all men of your contributions "for the benefit of mankind"' (Figure 26). The first such award, it could have had no other recipient than Niels Bohr. Hobnobbing with heads of state became routine (Figure 27).

In the year of this triumph, Bohr read the Danish translation of a book by a Swiss journalist, Robert Jungk, entitled, in its English edition, *Brighter than a thousand suns*. It is a romanticized tale of the atomic physicists who founded the science for the bomb and presided over creating on earth the manifestation of Krishna, 'brighter than a thousand suns', that Oppenheimer liked to quote from the Bhagavad Gita. Jungk's book featured Heisenberg's recollections of his visit to occupied Copenhagen in 1941. According to Jungk, Heisenberg told Bohr that although his group knew that ^{235}U and ^{239}Pu were fissionable, making an atomic weapon based on either of them would require so great and costly an effort that physicists could easily contrive to fail while appearing to their governments to be eager to succeed. Consequently, if Bohr could assure German physicists that their American and British counterparts would adopt the same tactics, the world would be a better place. Germany in particular would be spared a nuclear attack; for, according to Jungk, by the time of his visit to Copenhagen in 1941, Heisenberg had worked out that Germany would lose the war, and that, if either side had achieved atomic weapons before then, it would not be the Germans. But,

26. Bohr at the time he received the first Atoms for Peace Award.

still according to Jungk, Heisenberg expressed himself so guardedly that Bohr misunderstood him to say that Germany was proceeding apace to build an atomic bomb.

To this story, obtained by interview, the Danish edition of Jungk's book adds a letter from Heisenberg that contains significant

27. Bohr receiving Queen Elizabeth and, more relaxed, the Prime Minister of Israel.

detail. The conversation with Bohr 'probably began' with Heisenberg's asking whether physicists had the moral right to work on atomic problems during wartime. Bohr asked in reply ('as far as I can remember') whether Heisenberg thought that atomic weapons were possible; to which he answered that he knew they were, and repeated his first question. From this exchange Bohr derived the erroneous impression that Germany had made great progress in the pursuit of atomic weapons. Perhaps to improve its credibility, Heisenberg located this interview in time and place: 'on an evening walk in a district near Ny-Carlsberg', that is, around Bohr's villa.

These statements with their self-serving suggestion of high moral purpose shocked Bohr. He recalled that Heisenberg had come to Copenhagen in 1941 to try to persuade Danish scholars and intellectuals to cooperate with the Reich. Contrary to Jungk's rendition, Bohr remembered that Heisenberg and his colleague, Carl von Weizsäcker, a former visitor to the Institute, believed confidently in a German victory, and urged collaboration as the best course for their Danish friends. Bohr recalled further that his meeting with Heisenberg took place in his office in the Institute. The disagreement with Heisenberg's *mise-en-scène* is striking. Bohr's recollection must be preferred to Heisenberg's: meeting privately in or around Bohr's villa for so serious a conversation would have been compromising to both of them. Bohr's account of the interview, which comes from a draft of an undated unsent letter to Heisenberg probably written soon after Bohr read Jungk's book, concludes: '[I]n vague terms you spoke in a manner that could only give me the firm impression that, under your leadership, everything was being done in Germany to develop atomic weapons.'

It is difficult to know what Heisenberg wanted to tell Bohr. Perhaps he did not know himself. Did he believe that Germany would win? Then his effort to enlist Bohr in a holy band of physicists sworn to sabotage the development of a weapon that, if achieved first by the Allies, might save them was an invitation to treason. If, as seems very unlikely in the autumn of 1941, he foresaw a German defeat, his offer not to pursue atomic weapons would again be treason. And if, as Bohr conjectured, he was trying to discover what Bohr knew about parallel efforts in the USA and Britain, he was betraying a personal trust. Bohr: 'I listened to [you] without speaking since a great matter for mankind was at issue in which, despite our personal friendship, we had to be regarded as representatives of sides engaged in mortal combat.'

Heisenberg's exclusively German team worked vainly through the war to create a self-sustaining chain reaction (a 'radioactive pile') of the type that Fermi, continuing his international collaboration, achieved in Chicago in December 1942. At the war's end, the Allies rounded up the unsuccessful German team and sent them into pleasant detention in a country house in England. Their well-appointed quarters were well bugged. Believing themselves in private, they expressed great chagrin and dismay over the atomic bombing of Japan, for they had thought that they knew more about the processes involved than the Allies. Their expectation that they would have marketable expertise after the war perished in the news from Hiroshima.

Bohr was deeply disturbed by Heisenberg's attempt to enrol him in the farce of portraying the German uranium project as an ongoing programme of active resistance. The advocate of Atoms for Peace, the campaigner for an open world, wanted no such part. To guard against any suspicion that he did, he informed Heisenberg (in his unsent letter!) that he 'had thought it appropriate' to send copies of it to the Danish Foreign Office and the German ambassador, General Duckwitz, who had alerted the Danes to the Nazi plans for arresting the Jews in 1943. In later letters, sent and unsent, and in conversations during their later exchanges at professional meetings, Bohr pressed Heisenberg to reveal the authority that had given him permission to disclose the uranium project to an enemy alien. Heisenberg's reply, if any, does not seem to exist, leaving open the question whether, as he asserted, he acted at his own discretion with the advice of members of his team, or whether he was put up to fishing for information by his government.

Historia magistra vitae

Bohr's fellow countrymen trusted him to lead them along many complementary paths. Beside his ongoing presidencies or chairmanships of the Royal Danish Academy of Sciences and Letters, the Danish Atomic Energy Commission, and the Danish Cancer Society, and the directorship of his Institute, he took on transient responsibilities for several charitable organizations. One that deserves mention is the Bering Committee, set up to give medals and raise a monument in honour of Vitus Bering, the Danish explorer in Russian service immortalized in the Bering Strait. Bohr took the occasion of the 200th anniversary of Bering's death in December 1941 to encourage his fellow citizens: honouring the illustrious dead is 'a main source of invigoration for ourselves', of 'reassurance that precisely in our traditions we can find the strength…[we need] in the trying time through which we must now live'.

As repository of wisdom and guarantor of quality, Bohr wrote prefaces to books, for example, the Danish translations of Oppenheimer's *Science and the common understanding* (1953, 1960), and I. B. Cohen's *The birth of a new physics* (1960), which has nothing to do with the new physics Bohr made. Cohen's book, which deals with the 17th century, was a volume in a series intended to inculcate the culture as well as the technique of physical science. Bohr was a member of a small committee that brought the series, developed in the USA, to Denmark. In his last years Bohr took an increasingly strong interest in what his committee described as 'the spreading of information in the widest circles about the history and present state of physics'.

Bohr had acquired some information about the history of physics in the course of writing memorial addresses and obituary notices of such scattered notables as Galvani, Robert Boscovich, Ørsted, James Clerk Maxwell, and Janne Rydberg (after whose name and work the 'R' in the Balmer formula comes), and refreshed his knowledge of its almost present state with accounts of Pieter Zeeman, Rutherford, Einstein, Pauli, and Heisenberg. Although most are flimsy, occasional pieces, several suggest considerable background preparation and those on Ørsted, Rutherford, and Einstein add to the historical literature. Bohr's lecture on the centennial of Ørsted's death in 1951 is almost autobiographical. We learn about the close intellectual tie between Ørsted and his brother; the dependence of a great discovery (in Ørsted's case electromagnetism) on a deep (though obscurely expressed) philosophy of nature; the great joy in seeing one's initiative followed up all over Europe; the importance of international collaboration in the advance of science; efforts to improve science education and the conditions of scientific work in Denmark; foundation or establishment of institutions (in Ørsted's case the Polytechnic, the Academy of Sciences and Letters, the Society for the Dissemination of Natural Science); in short, the combination of a profound thinker and committed humanist. It is no surprise to learn that Bohr's epistemology justified Ørsted's: 'we can precisely on the new background appreciate deeply his characteristic striving to unite what he called the beautiful and the true'.

Bohr's main source for Ørsted's life and works was Kirstine Meyer, a great friend of his Aunt Hanna, and like her a teacher trained in physics. Meyer came to specialize in its history and earned a doctor's degree from the University of Copenhagen (the first for a woman in physics) when almost 50 with a thesis on the history of the concept of temperature. Meyer subsequently became the leading authority on Ørsted, whose papers she edited and published in three volumes in 1920, discovering in the process detailed notes on the many experiments that lay behind the compressed Latin account of his great discovery. Bohr read the proofs and noticed the contrast between the preparatory details and the compressed presentation, which had unfortunate parallels in his own writing. A second source of Bohr's approach to Ørsted was the careful study he had made to compose a preface to an eight-volume work, *Danmarks kultur ved aar 1940*, published between 1941 and 1943 to recall the country's heritage during the German occupation. Bohr's concept of Danish culture, like his analysis of its exemplar Ørsted, had a measure of autobiography as well as of patriotism. Its prime characteristics to him were its ability to graft foreign elements onto Danish roots and to welcome sound cultural contributions, and contributors, from everywhere. Great Danish men of science have gone abroad, learned, and returned to plant worthwhile foreign practices in Danish soil: witness Tycho Brahe, Ole Rømer, Ørsted, and Bohr himself. Danes have not only welcomed but enticed foreign scholars: Bohr instanced the Rask–Ørsted Foundation, whose fellowships helped to make his Institute an international centre between the wars.

How did so small a country manage to take continually from such powerful neighbours as Britain and Germany without losing its identity? Until modern times its remoteness and its distinct language insulated it; foreign ideas sank in slowly enough to be assimilated into the national culture. During the 19th century, an enlightened state freed the peasants, introduced a democratic constitution (which, incidentally, by removing all civil disabilities against Jews brought Bohr's grandfather back to Denmark from England), and established novel institutions to educate its small population in modern subjects and Danish culture. They were thus prepared to meet and greet everything of value, to preserve that 'attitude to the fellowship among peoples that our entire history has fostered [and] might well be the most characteristic feature of our culture... No matter how far-reaching the consequences, in all areas of human life... we are entitled to hope that our people, as long as we can retain freedom to develop the outlook that is so deeply rooted in us, will also in the future be able to serve honourably the cause of mankind.'

Although the situation had a complementary negative side, the Danish social-democratic government that accepted cooperation with the German occupiers contrived to ensure the survival of a liberal, democratic, open Denmark. During the 1930s it had passed reforms to improve social cohesiveness, combated anti-Semitism, and pressed the population to accept democratic ideals as integral to Danish culture. It held the Danish Nazi Party to 1 per cent of the vote in the general election of 1935. And in the elections of 1943, though held during the occupation, it kept the party below 2 per cent.

An invitation by the Physical Society of London to give its Rutherford Memorial Lecture in 1958 returned Bohr to the complexities of historical reconstruction he had had to face in responding to Jungk's report of Heisenberg's visit. Preparing the lecture prompted a more serious review of the genesis of the Trilogy and subsequent developments than he had thought necessary for his many previous synopses of atomic physics during the decade preceding quantum mechanics. He now insisted on including errors and false starts, an approach as unpalatable to a physicist as renunciation of the conservation of energy. Thus Darwin, to whom he sent a draft of the greatly expanded version of his talk finally published in 1961, objected that Bohr unnecessarily cluttered the narrative, 'with all of the difficulties being brought out all the time', rather than giving a clear and logical overview of his discoveries about the hydrogen atom. Bohr replied after much deliberation that he could not follow Darwin's suggestion without doing violence to history. 'I have striven to use the opportunity to revive the development in a factual and detached manner.' He knew that he had not begun his analysis with hydrogen or even with spectroscopy. That is what can be found in textbooks. Bohr insisted on beginning where he began, and to put it all down. 'It has been quite a difficult task, and I have often been scared by the length of the manuscript, resulting from the many different points that presented themselves to give the whole story a reasonable balance.'

Bohr's last prepared historical work was a lecture on the history of the Solvay conferences on physics delivered in October 1961 on the occasion of the fiftieth anniversary of the first one. He said, rightly, that the reports of the conferences are a most valuable source for students of the history of science; more reliable, indeed, than the account he there gave of the origin of his atomic model. For then he followed Darwin's advice, though in a manner that allowed for the complexities in the Rutherford lecture. 'A starting point was offered by...the optical spectra of the elements': true, but it was not Bohr's starting point! He did, however, take the trouble to read the reports of the first two Solvay meetings, and perhaps of the others too, with great care. He learned for the first time that the experts on radiation and the quantum who gathered in Brussels in 1911 did not once mention the nuclear atom. Nor did Rutherford's invention receive much attention at the second meeting, in October 1913, which was devoted to atomic structure; while Bohr's first paper on the subject, which had appeared in June that year, did not enter the discussions at all. Like other people, physicists sometimes lose the compass they need to find their way. 'It is illuminating for the understanding of the general attitude of the physicists at that time [Bohr remarked, in a sentence without a compass] that the uniqueness of the fundament for such exploration given by Rutherford's discovery of the atomic nucleus was not yet generally appreciated.'

Heisenberg was among the Solvay celebrants in 1961. Bohr thought to enrol him in the new set of problems. 'In recent years I have become increasingly engaged in historical questions...dwelling more and more on thoughts of the great adventure which we all experienced.' Won't you help? Bohr confided that an American team sponsored by the 'Academy in Washington' (in fact by the US National Science Foundation) proposed to interview him at length and many enquirers had asked for archival material concerning the 'preparation and discussion' of uranium projects during the war. The more he tried to peel back the layers, the more he experienced a problem more challenging than measurement in quantum theory. No longer was it a question of a single observer and the observed: 'the difficulty [lies in] giving an accurate account of developments in which many different people have taken part'.

In the summer of 1962, the American team led by Thomas Kuhn left Berkeley for a year's stay in the converted stables attached to the Carlsberg villa. Although Bohr died a few weeks after interviewing began, his extensive archive of professional correspondence and collection of reprints provided material that helped the team to prepare oral histories with others, including Heisenberg, and in persuading them to lend their own archives

28. A typical international conference at the Institute, 1936, with major actors in the front row: Pauli, Jordan, Heisenberg, Born, Meitner, Otto Stern, Franck, Hevesy. Others present mentioned in this story are von Weizsäcker and Frisch (row 2, 4th from left and 2nd from right); Delbrück and Kramers (row 5, 3rd and 6th from left); Bohr and Rosenfeld are standing together on the left. Conferences on history of physics, equal in number if not in brain power of participants, now take place in the same venue.

for microfilming. The Niels Bohr Archive in Copenhagen holds abundant documentation of the great adventure Bohr led. In continuance of his programme and Danish tradition, an international group of scholars continues to study the adventure and developments arising from it (Figure 28).

Bohr received awards and distinctions, medals and prizes, in such numbers as to attract the attention of cartoonists (Figure 29). Perhaps the most notable of his honours, because domestic and reserved for royal and similarly exalted personages, was the Order

29. Bohr laden with honours introduced to lecture on 'chain reactions'.

of the Elephant. For this a coat of arms was required. Bohr created one that can serve as his epitaph. A shield at its centre carries the charge of the Chinese symbol of yin and yang, the composition of opposites, under a scroll reading, in Latin, *contraria sunt complementa.* Encapsulating his thought in the ancient language makes it appropriately oracular.

References

Full citations for items in Further reading will be found there.
AH signifies Aaserud and Heilbron, *Love, literature, and the quantum atom*; *CW*, Bohr, *Collected works*; NBA, Niels Bohr Archive, Niels Bohr Institute, Copenhagen.

Prologue

'liberate himself': *CW, 10*, 194 (text of 1957).

'a vote': Sir Henry Dale to Winston Churchill, 11 May 1944, in *CW, 11*, 241.

'Tireless searching': as translated in Edwin H. Zeydel, *Goethe the lyricist* (Chapel Hill: U. North Carolina P., 1955), 171.

Chapter 1: A richly furnished mind

'special, very concrete': Lorenz Rerup, 'Judisk indsats i dansk vetenskab', in Harald Jørgensen, ed., *Indenfor murene, Judisk liv i Danmark* (Viborg: Reizel, 1984), 215.

'I too could think': Bohr to Margrethe, 21 Dec. 1911, in AH, 161.

'intimate special life', 'wily, equivocal', 'world of the mind', 'special exclusivity': Henri Nathansen, *Jude oder Europäer? Portrait von Georg Brandes* (Frankfurt/Main: Rütten and Loening, 1931), 46, 42, 50, resp.

'so infinitely trivial': Bohr to Sophie Nørlund, 1 May 1912, in AH, 77.

'When I see', 'tell me whether': Bohr to Margrethe, 17 Dec. 1911, in AH, 43.

'I will come to you', 'Dear Niels', 'I set no limit': Margrethe to Bohr, 20 Oct. 1911, in AH, 157–8.

'after his lonely battles', 'could not imagine', 'his courage', 'My own little darling': Bohr to Margrethe, 21 Dec. 1911, in AH, 47.

'all the debts': AH, 160.

'rare treasure', 'gold', 'greatest and wisest': AH, 134, 127, 12.

'the nature': Harald Høffding, 'Philosophy and life', *International journal of ethics*, 12:2 (1902), 137.

'this state of affairs', 'an essential condition': Edgar Rubin, *Experimenta psychologica* (Copenhagen: Munksgaard, 1949), 20.

'by the ideals': Høffding, *Søren Kierkegaard als Philosoph* (Stuttgart: F. Frommann, 1896), 3–4.

'Even one': Høffding, *The problems of philosophy* (New York: Macmillan, 1905), 180.

'in the goodness' to 'the good': Bohr to Sophie Nørlund, 1 May 1912, AH, 77.

'He to whom': Høffding, *Philosophy* (1906), 3.

'for whom his lectures': Erik Rindom, *Harald Høffding* (Copenhagen: Gyldendal, 1913), 70, 79.

'the most valuable thing I possess': Bohr to Margrethe, 16 July 1912, in AH, 92.

'good friend', 'Niels Bohr': Høffding to Meyerson, 12 Feb. 1924 and 13 Apr. 1928, in Fritiof Brandt et al., eds, *Correspondance entre Harald Høffding et Emil Meyerson* (Copenhagen: Munksgaard, 1939), 70, 156.

'ideas that helped': Høffding to Emil Meyerson, 20 May 1923, ibid., 51, re Høffding, *Der Begriff der Analogie* (1924); and 30 Mar. 1928, ibid., 149, re Bohr's éloge on Høffding's 85th birthday (*CW*, *10*, 308–9).

'the great question', 'an irrational relationship', 'for us existence': Høffding, *Problems* (1905), 93–4, 107.

'Here again': ibid., 113.

'an exhaustive concept': ibid., 114–15.

'the irrationality': Høffding, *International journal of ethics*, 22:2 (1902), 149.

'we live forward': Høffding, *Kierkegaard* (1896), 2.

'A situation': *CW*, *10*, 200.

'I walk here': Bohr to Harald Bohr, 20 Apr. 1909, in *CW*, *1*, 501.

'made a powerful impression': remarks by Bohr in 1933 recorded by J. Rud Nielsen as quoted by Gerald Holton, 'The rooots of complementarity', *Daedalus* (Fall 1970), 1053 n. 47.

'It is the only thing I have to send': Bohr to Harald, 20 Apr. 1909, *CW*, *1*, 501.
'In love', 'romantic soul', 'modern philosophy,' 'He became': Kierkegaard, *Johannes Climacus*, tr. T. H. Croxall (Stanford, Calif.: Stanford UP, 1958), 103, 116, 140, 115.
'it could be': Bohr to Harald, 19 June 1912, in *CW*, *2*, 559.
'This seems to be': 'Rutherford Memorandum', in *CW*, *2*, 136–58, on 136.

Chapter 2: Productive ambiguity

'I found myself rejoicing': in AH, 135–6.
'A little crazy about': Bohr to Harald, 29 Jan. 1912, in AH, 152.
'leaving reality', 'the history of discovery': Joseph Larmor, *Aether and matter* (Cambridge: Cambridge UP, 1900), vi, ix, xiii.
'When I read': Bohr to Margrethe, 22 Feb. 1912, in AH, 114.
'Attempt to apprehend': Larmor, *Aether and matter*, ix.
'this seems to be': Manchester Memorandum, in *CW*, *2*, 136.
'I do not deal', 'Have you looked': Bohr, 'Interview' (Oct.–Nov. 1962) by T. S. Kuhn et al. (NBA).
'He was extremely astonished': Hevesy to Bohr, 23 Sept. 1913, in *CW*, *2*, 532,
'While there obviously': *CW*, *2*, 175.
'quite essential' to 'Most of it': Bohr, 'Interview', 61, 8, 57–9.
'Shall we': Bohr to Margrethe, 2 Apr. 1913, in AH, 97.
'there is a greater tendency', and quotations in the following paragraph: *CW*, *2*, 207, 207, 209, 214.
'For the present': Moseley to Bohr, 16 Nov., and draft reply, in Margrethe's hand, 21 Nov. 1913 (quote), in *CW*, *2*, 544–7.

Chapter 3: Magic wand

'the most faithful friend': Harald Bohr to Bohr, 30 Jan. 1912, in AH, 69.
'the mechanism': *CW*, *2*, 449.
'periods of overhappiness': Bohr to Owen Richardson, 15 Aug. 1918, in *CW*, *3*, 14.
'an international place': quoted in Peter Robertson, *The early years*, 34–5 (text of 26 Oct. 1919).
'different human conditions': ibid., 65.
'undeserved good fortune': ibid., 65 (text of 1922).
'we do not yet': Bohr to Coster, 10 Dec. 1924, in *CW*, *4*, 679–81.

'the difficulties': Bohr to R. H. Fowler, 5 Dec. 1924 (NBA).
'every description': Bohr, *CW, 3*, 458–9 (text of 1923).
'I am very depressed': Pauli to Alfred Landé, 23 May 1923, in Wolfgang Pauli, *Wissenschaftlicher Briefwechsel*, vol. 1, ed. Armin Hermann et al. (Berlin: Springer, 1979), 87.
'No one understands': Heisenberg to Pauli, 9 Oct. 1923, ibid., 125–6.
'It was very good': Pauli to Bohr, 11 Feb. 1924, ibid., 143–4.
'Purely formal', to 'Unphilosophical': collected from Pauli's letters to Alfred Landé, Kramers, and Bohr, Dec. 1623–Feb. 1624, ibid., 134–6, 143–4.
'I believe that': Pauli to Bohr, 12 Dec. 1924, ibid., 188.
'There is nothing else': Bohr to Fowler, 21 Apr. 1925, in *CW, 5*, 82.
'The thing': Slater, interview with T. S. Kuhn, 3 Oct. 1963, p. 34 (NBA).
'The magical': John C. Slater, *Solid state and nuclear theory: A scientific autobiography* (New York: Wiley, 1975), 42.

Chapter 4: Enthusiastic resignation

'such quantities', 'further reminiscence', 'it does not mean': H. A. Kramers, *Nature*, 114 (1924), 310–11, dated 22 July.
'*Mist*': Heisenberg to Pauli, 8 June and 28 July 1926, in Pauli, *Briefwechsel, 1*, 328, 338.
'monstrous': Schrödinger to H. A. Lorentz, 6 June 1626, in Walter Moore, *Schrödinger* (Cambridge UP, 1989), 221.
'The idea of your work': Einstein to Schrödinger, 16 Apr. 1926, ibid., 221.
'The discussions', 'But I think', 'formidable': Bohr to Fowler, 26 Oct. 1926, in *CW, 6*, 423–4.
'remarkable belief': Schrödinger to Willy Wien, 26 Aug. 1926, in Moore, *Schrödinger*, 228.
'the great Niels Bohr', 'as it seems to me', 'resting place', 'Certainly we can', 'What I have': Schrödinger to Bohr, 23 Oct. 1926, in *CW, 6*, 459–60; the '...' are in the original text.
'it is just': all quotations in this and the following five paragraphs come from Bohr, 'The quantum postulate and the recent development of atomic theory', *Nature*, 112: Suppl. (1928), 580–90, in *CW, 6*, 148–58.
'At first': all quotations in this and the following paragraph come from Ehrenfest to Samuel Goudsmit et al., 3 Nov. 1927, in *CW, 6*, 37–40, slightly revised using the German text, ibid., 415–18.

'at last': Oseen to Bohr, 8 July 1935, in Heilbron, 'Missionaries', 211.
'Physical reality': Franck to Bohr, 9 Jan. 1936, ibid., 212.

Chapter 5: The Institute

'Just as we': *CW*, *9*, 130 (text of 1923).
'the invisible': Heilbron, 'Missionaries', 225 (Jung, letter of 1956).
'one must get used to [them]': Jordan, ibid., 216.
'better than any': Léon Rosenfeld, *Collected papers*, ed. R. S. Cohen and J. J. Stachel (Dordrecht: Reidel, 1979), 535.
'So many': Bohr to Kramers, 14 Mar. 1936 (NBA).
'renunciation of mechanical models': Bohr, *Atomic theory and the description of nature* (Cambridge: Cambridge UP, 1934), 50.
'the general conditions': the quotations in this and the following paragraph, ibid., 15, 93, 96, 100–1.
'Resignation': quotations in this paragraph, ibid., 115, 116, 119.
'The very existence': Bohr, *Atomic physics and human knowledge* (New York: Wiley, 1958), 9.
'basic postulate', 'biological regularities': ibid., 21.
'gradual removal': ibid., 31.
'the conviction': *CW*, *12*, 423.

Chapter 6: Elder statesman

'Distinguished son': *The New York Times*, 1 Nov. 1954, in *CW*, *11*, 370.
'the importance': *CW*, *11*, 641.
'Probably began', 'as far as I can remember', 'on an evening': Robert Jungk, *Brighter than a thousand suns* (Harmondsworth: Penguin, 1964), 100–1 (German original, 1956).
'In vague terms': Dörries, *Copenhagen*, 109 (text of *c.*1957).
'I listened': ibid., 109–10.
'He had thought it': ibid, 113.
'a main source': *CW*, *12*, 244–5.
'spreading of information': ibid., 95.
'we can precisely': *CW*, *10*, 342.
'attitude to the fellowship': ibid., 272.
'with all of the difficulties': Darwin to Bohr, 11 Sept. 1961, in *CW*, *10*, 463.
'I have striven', 'It has been quite': Bohr to Darwin, 20 Sept. 1961, in *CW*, *10*, 464.

'A starting point': *CW, 10*, 436.
'It is illuminating': ibid.
'In recent years': quotations for the paragraph, Dörries, *Copenhagen*, 119, 125, 135, 149, 169 (letters Bohr drafted in November and December 1961 and March 1962, none of which seems to have been sent).

Further reading

Aaserud, Finn. *Redirecting science: Niels Bohr, philanthropy and the rise of nuclear physics*. Cambridge: CUP, 1990.

Aaserud, Finn, and J. L. Heilbron. *Love, literature and the quantum atom: Niels Bohr's 1913 trilogy revisited*. Oxford: OUP, 2013.

Bohr, Niels. *Atomic theory and the description of nature*. Cambridge: CUP, 1934, reprinted 1961.

Bohr, Niels. *Atomic physics and human knowledge*. New York: Wiley, 1958.

Bohr, Niels. *Collected works*. Ed. Léon Rosenfeld et al. 12 vols. Amsterdam: North-Holland, 1972–2007.

Dörries, Matthias. *Michael Frayn's Copenhagen in debate: Historical essays and documents on the 1941 meeting between Niels Bohr and Werner Heisenberg*. Berkeley: University of California, Office for the History of Science and Technology, 2005.

Folse, Henry J. *The philosophy of Niels Bohr: The framework of complementarity*. Amsterdam: North-Holland, 1985.

Heilbron, J. L. 'The earliest missionaries of the Copenhagen spirit', in Edna Ullmann-Margalit, ed., *Science in reflection*. Dordrecht: Kluwer, 1988. Pp. 201–33.

Kragh, Helge. *Niels Bohr and the quantum atom: The Bohr model of atomic structure 1913–1925*. Oxford: OUP, 2012.

Lidegaard, Bo. *A short history of Denmark in the 20th century*. Copenhagen: Gyldendal, 2009.

Pais, Abraham. *Niels Bohr's times, in physics, philosophy, and polity*. Oxford: OUP, 1991.

Robertson, Peter. *The early years: The Niels Bohr Institute 1921–30*. Copenhagen: Akademisk Forlag, 1979.
Rozental, Stefan, ed. *Niels Bohr: His life and work as seen by his friends and colleagues*. Amsterdam: North-Holland, 1968.

Index

Note: To meet the needs of digital users, the Index was compiled using a program that may occasionally result in a reference given as spanning two pages being found on only one.

C

D

I

J

K

L

M

N

O

P

Q

R

S

T

U

V

W

X

Z

Forensic Science
A Very Short Introduction
Jim Fraser

In this Very Short Introduction, Jim Fraser introduces the concept of forensic science and explains how it is used in the investigation of crime. He begins at the crime scene itself, explaining the principles and processes of crime scene management. He explores how forensic scientists work; from the reconstruction of events to laboratory examinations. He considers the techniques they use, such as fingerprinting, and goes on to highlight the immense impact DNA profiling has had. Providing examples from forensic science cases in the UK, US, and other countries, he considers the techniques and challenges faced around the world.

> An admirable alternative to the 'CSI' science fiction juggernaut... Fascinating.
>
> William Darragh, Fortean Times

GENIUS

A Very Short Introduction

Andrew Robinson

Genius is highly individual and unique, of course, yet it shares a compelling, inevitable quality for professionals and the general public alike. Darwin's ideas are still required reading for every working biologist; they continue to generate fresh thinking and experiments around the world. So do Einstein's theories among physicists. Shakespeare's plays and Mozart's melodies and harmonies continue to move people in languages and cultures far removed from their native England and Austria. Contemporary 'geniuses' may come and go, but the idea of genius will not let go of us. Genius is the name we give to a quality of work that transcends fashion, celebrity, fame, and reputation: the opposite of a period piece. Somehow, genius abolishes both the time and the place of its origin.

www.oup.com/vsi

KEYNES

A Very Short Introduction

Robert Skidelsky

John Maynard Keynes (1883–1946) is a central thinker of the twentieth century, not just an economic theorist and statesman, but also in economics, philosophy, politics, and culture. In this *Very Short Introduction* Lord Skidelsky, a renowned biographer of Keynes, explores his ethical and practical philosophy, his monetary thought, and provides an insight into his life and works. In the recent financial crisis Keynes's theories have become more timely than ever, and remain at the centre of political and economic discussion. With a look at his major works and his contribution to twentieth-century economic thought, Skidelsky considers Keynes's legacy on today's society.

www.oup.com/vsi

GERMAN PHILOSOPHY

A Very Short Introduction

Andrew Bowie

German Philosophy: A Very Short Introduction discusses the idea that German philosophy forms one of the most revealing responses to the problems of 'modernity'. The rise of the modern natural sciences and the related decline of religion raises a series of questions, which recur throughout German philosophy, concerning the relationships between knowledge and faith, reason and emotion, and scientific, ethical, and artistic ways of seeing the world. There are also many significant philosophers who are generally neglected in most existing English-language treatments of German philosophy, which tend to concentrate on the canonical figures. This *Very Short Introduction* will include reference to these thinkers and suggests how they can be used to question more familiar German philosophical thought.

www.oup.com/vsi

HERODOTUS

A Very Short Introduction

Jennifer T. Roberts

Herodotus: A Very Short Introduction introduces readers to what little is known of Herodotus' life and goes on to discuss all aspects of his work, including his fascination with his origins; his travels; his view of the world in relation to boundaries and their transgressions; and his interest in seeing the world and learning about non-Greek civilizations. We also explore the recurring themes of his work, his beliefs in dreams, oracles, and omens, the prominence of women in his work, and his account of the battles of the Persian Wars.

MUHAMMAD
A Very Short Introduction
Jonathan A. C. Brown

As the founder of Islam Muhammad is one of the most influential figures in history. The furore surrounding the Satanic Verses and the Danish cartoon crisis reminded the world of the tremendous importance of the prophet of Islam, Muhammad. Learning about his life and understanding its importance, however, has always proven difficult. Our knowledge of Muhammad comes from the biography of him written by his followers, but Western historians have questioned the reliability of this story in their quest to uncover the 'historical Muhammad'. This *Very Short Introduction* provides an introduction to the major aspects of Muhammad's life and its importance, providing both the Muslim and Western historical perspectives.

'This is an excellent introduction to the life of Muhammad. Dr Brown is providing the reader with a rigorous study based on the classical Islamic tradition, yet well balanced between elements of faith and rational discussions, useful for Muslims and non Muslims alike. Very easy to read, profound and interesting to study.'

Tariq Ramadan

www.oup.com/vsi